"Smart cookie, that Jen Lin-Liu. [She] knows the world loves a good cook, especially one who feeds a hunger for Chinese culture."
— *Washington Post*

"A captivating, often humorous tale of the Chinese-American author's stint in cooking school in Beijing, her time shaving noodles in a self-serve eatery, and her work in a posh restaurant in Shanghai." — *Newsday*

"Lin-Liu is the sort of cook who writes without embarrassment of 'the beauty of noodles' and 'the power of dumplings.' [A] lively account of learning to cook in China." — *Christian Science Monitor*

"[Lin-Liu] charts her journey of discovery engagingly... an enticing taste of Chinese food and cooking culture." — *Atlanta Journal-Constitution*

"Her writing is vivid and fun; her stories give us a remarkable and intimate look at everyday life in China." — *Minneapolis Star Tribune*

"Entertaining and offbeat." — *Fresh Air*

"When American Lin-Liu enrolls in a cooking school in Beijing, her goals are modest: But after she interns at a noodle stall and masters dumplings, the intrepid author gets hooked on a culinary scene where both foie gras and animal genitalia are on offer." — *People*

"[A] lively and rare look at Chinese cuisine." — *Pittsburgh Post-Gazette*

"Entertaining and eye-opening, Lin-Liu's portrait of modern China reflects its changing trends and attitudes and its timeless cuisine."
— *Portsmouth Herald*

"*Serve the People* leaves the pretension at the kitchen door and makes your stomach rumble in anticipation for dinner." — *Time Out Beijing*

"A Joy Luck Club of Chinese cuisine... a surprisingly addictive page-turner." — *Far Eastern Economic Review*

"[An] entertaining read... punctuated with fascinating social commentary, not to mention mouthwatering descriptions of food, authentic recipes, and lighthearted anecdotes." — *Northwest Asian Weekly*

SERVE THE PEOPLE

SERVE THE PEOPLE

A Stir-Fried Journey Through China

JEN LIN-LIU

Mariner Books
Houghton Mifflin Harcourt
Boston New York

For my parents

First Mariner Books edition 2009

Copyright 2008 by Jen Lin-Liu

All Rights Reserved

For information about permission to reproduce selections from this book, write to Permissions, Houghton Mifflin Harcourt Publishing Company, 6277 Sea Harbor Drive, Orlando, Florida 32887-6777.

www.hmhbooks.com

Library of Congress Cataloging-in-Publication Data
Lin-Liu, Jen.
Serve the people: a stir-fried journey through China. / Jen Lin-Liu.—1st ed.
   p. cm.
   1. Cookery, Chinese. 2. Cookery—China. 3. Lin-Liu, Jen. 4. China—Social life and customs. I. Title.
TX724.5.C5L5555 2008
641.5951—dc22
2008008155
ISBN 978-0-15-101291-6
ISBN 978-0-15-603374-9 (pbk)

Printed in the United States of America
DOC 10 9 8 7 6 5 4 3 2 1

Book design by Lisa Diercks
Illustrations by Kheng Guan Toh/iStock
Typeset in Manticore, Goudy Sans, and Weiss Initials

Contents

List of Recipes

part one

COOKING SCHOOL

1

In cooking class, I learned a startling array of things: Eating fish head will repair your brain cells. Spicy food is good for your complexion. Monosodium glutamate is best thrown in a dish just before it comes off the wok. Americans are fat because they eat bread, while Chinese are slim because they eat rice. If you work as a cook in America for three years, you can come back to China and buy a house.

I bicycled down a narrow alley, past a public toilet and through a gate with a scowling security guard to get to the cooking school. I enrolled in October 2005, in my fifth year of living in China. Lectures were held in a classroom rented from a high school, and the unheated room felt colder by the minute. Everyone else was used to the lack of heat and insulation in public schools and dressed accordingly, in down jackets. I shivered in my thin coat.

My classmates slumped in their seats, seeming bored and detached, holding their pens limply. They were all men, ranging in age from twenty to fifty. Most hadn't completed high school. Teacher Zhang didn't mind that they answered their cell phones in class. Once I heard a student clipping his fingernails, the snip of the scissors punctuating the cadences of the lecture. Teacher Zhang often narrowed his eyes at me while he spoke, however. He didn't like that I was different from the others.

I interrupted him with questions. I didn't bother to raise my hand because that custom didn't exist in this classroom — students weren't supposed to have questions. So I just spoke up, as loudly as I could. "Could you write that character more clearly?" I often asked.

Teacher Zhang grunted and wiped the chalkboard with the back of his hand, then rewrote the character in the Chinese equivalent of block letters, the chalk screeching against the surface.

Sometimes I was so busy copying down what Teacher Zhang said that I didn't stop to question it. Later, reviewing my notes, I discovered a strange mix of pop psychology and old Chinese clichés.

"Your taste buds are influenced by several different factors," Teacher Zhang declared, and enumerated them. *Age, sex, occupation, and mental state,* I dutifully wrote. Women liked lighter foods and men preferred spicy foods, he explained. *Women–light, men–spicy,* I scribbled in English.

"If you're a farmer working in the fields and doing a lot of heavy lifting, your taste buds will be different from someone who works in an office," he added. Uh-huh, I thought as I transcribed the information into my notebook. Looking back on my notes later, I wished I had asked him to elaborate further.

But when I asked questions, Teacher Zhang shot me annoyed glances, and the other students shifted uncomfortably in their seats. I learned to behave like the rest: listen, bow, and copy.

. . .

When we weren't in the classroom taking notes on the world according to Teacher Zhang, we were in the kitchen. The kitchen was created out of another classroom by installing a burner, gas tank, countertop, sink, and refrigerator. With those simple fixtures, it became a larger version of the typical home kitchen in China. Though spacious, it wasn't the kind of room where you let your eyes wander, lest they settle on a patch of scum on the tile wall that had probably not been scrubbed for a good five years. I figured the time we would spend in the kitchen learning real things would make up for the time wasted on the lectures. I had never been in a professional Chinese kitchen, which was notoriously off-limits to diners. I now found myself in a room full of cutting boards, woks, cleavers, and bottles of chili oil and oyster sauce. But I quickly found out that even in this kitchen, certain things were not permitted to students. Like cooking.

Instead of cooking, we sat on a set of bleachers across the room, observing Chef Gao's every move. Gao was an old-school chef who worked at a Soviet-style, government-run hotel that had seen its glory days pass with the end of the state-planned economy. Gao continued the time-honored Chinese traditions of using MSG and copious amounts of soybean oil. Despite the MSG and the oil—or maybe because of them—his cooking was delicious.

We watched as Chef Gao fanned out his skinny arms and elbows like a grasshopper as he cooked. We listened to him chirp the recipes in a folksy Beijing accent as he wrote them on the chalkboard, dividing the ingredients into three categories: main ingredients, supplementary ingredients, and seasonings. Occasionally, he noted quantities next to the main ingredients, but usually he threw things in by intuition. In any case, the kitchen didn't contain a single measuring cup or spoon.

We scrutinized his equipment, which was limited to a wok, a cutting board, and a cleaver with an eight-inch-long and four-inch-tall blade. Once in a while he'd pull out something more sophisticated, like a fryer basket. "You see this basket and handle? It's all one piece, so it will never break. I bought this in the 1960s, and the factory has been closed for years. They don't make baskets like this anymore!" he hollered with a disdain that suggested that the old days, when China was dirt-poor, were better.

We gazed, captivated, as Chef Gao demonstrated his skills at the wok. He lined the curved pan with thin slices of pork tenderloin that had been marinated in rice wine and placed it over a moderately high flame, swirling the juices around. When one side was cooked, he picked up the wok and flipped the pork in a single sheet, like a pancake. He repeated the maneuver — swirl, flip, swirl. Seasoned with leeks and ginger, this tender and flavorful *guota liji* (pan-fried pork tenderloin) displayed the fundamentals of Chinese cooking: freshness and simplicity above all else.

When all the day's dishes were cooked, we went into action. We jumped off the bleachers, gathered around the cooking station, and swooped in with chopsticks that we had brought from home, attacking the food with a unified strategy. The smallest dishes went first, especially expensive items like seafood. We moved on to the dish giving off the most steam, and then we finished the rest. In three minutes flat, everything was gone. At the end of one class, I barely beat someone to the last skewer of deep-fried scallops. He ended up with an empty stick as the scallops slid off into my greedy grasp. I had already learned that I couldn't stand around waiting for anything to be handed to me.

PAN-FRIED PORK TENDERLOIN (*GUOTA LIJI*)

¾ pound pork tenderloin. thinly sliced against
  the grain
2 tablespoons rice wine or sherry
½ teaspoon salt
½ teaspoon freshly ground white pepper
2 large eggs
½ cup all-purpose flour
¼ cup vegetable oil plus 1 tablespoon for drizzling
2 tablespoons chicken stock
1 leek, white part only, cut in half lengthwise
  and shredded
2 thumb-sized pieces of ginger, peeled and shredded
2 teaspoons sesame oil

Marinate the pork in 1 tablespoon of the rice wine, ¼ teaspoon of salt, and ¼ teaspoon of pepper for 10 minutes. In a bowl, beat the eggs and set them next to the stove.

Place the flour on a plate. When the pork has marinated, coat each slice with flour on both sides, patting to remove the excess. Set the slices on a plate next to the eggs.

Place a wok over medium heat and add ¼ cup of oil, swirling it to coat the sides. When the oil is hot, quickly dip each piece of pork in the beaten egg and place it in the wok, arranging the slices so they cover the bottom and sides in a thin sheet. With a spatula, gently loosen the pork, then drizzle a little oil around the wok from time to time so the meat doesn't stick as it cooks. When the bottom of the pork sheet has turned a light golden brown, flip it over. (Don't worry if it doesn't flip in a single sheet; just make sure to turn over each piece.) Add any remaining oil, the remaining 1 tablespoon of wine, the rest of the salt and pepper, and the chicken stock.

Sprinkle the shredded leek and ginger over the pork, and drizzle sesame oil over all. When the second side of the pork is browned, remove the wok from the heat and slide the pork onto a plate. Serve immediately.

The Hualian Cooking School was one of 129,000 hits that had come up when I Googled "Beijing cooking school" in Chinese. I chose the school mainly because it was in my central Beijing neighborhood—a big factor considering the capital's monstrous size and horrendous traffic. I was looking for a typical experience. I didn't start out with an ambitious goal; I figured I would be happy if I could become reasonably adept in Chinese cooking, good enough to hold a decent dinner party.

Like hundreds of other cooking schools in Beijing, Hualian was vocational: it aimed to train people who wanted to go into the restaurant industry. Students constantly drifted in and out of the classes, much the way people drifted in and out of jobs in many parts of China. Courses were taught on a rolling basis, and the school allowed newcomers to sit in on a class or two without paying. When I called to inquire about the cooking school's programs, the receptionist breezily said, "I'd recommend that you sign up for the intermediate class." How did she know if I qualified for something intermediate? I wondered. I found out when I got to the school: the intermediate-level course was the only one offered. The three-month program met Monday through Friday for two hours each day. A core group of ten of us began and finished the classes around the same time, just before a national cooking exam, held several times a year, was scheduled. After passing the exam, the students would receive a diploma and could proceed to find a job.

Our course focused on the main cuisines of China: the northern coastal Lu, Huaiyang from the Shanghai region, the southern or Cantonese style called Yue, and spicy Sichuanese from the country's interior, known simply as Chuan. These regional cuisines were designated the "four big" cuisines sometime in the late Qing Dynasty, though no food expert or chef ever gave me a satisfactory answer as to why people still used those four regions to define Chinese cuisine. It didn't reflect what Chinese were eating today. Rarely did I go out for Lu cuisine; fiery Hunanese and Thai-influenced Yunnanese cooking, which weren't on the list, were more popular among my Chinese peers than Huaiyang cuisine, which the Chinese said catered to the elderly with its lighter, more natural flavors. (But when I raised these points in class, the teacher inevitably gave me the stare. I had forgotten: my role was to listen, bow, and copy.)

The classes seemed inexpensive to me, but the other students carefully weighed the cost and the payoff before signing on. Two months' worth of classes cost $100. Upon graduating, a chef could expect to make $150 a month at an average restaurant, and perhaps, after a couple of decades of climbing the ladder like our teachers, could top out at around $500 a month, a comfortable salary by Chinese standards. Being a chef wasn't a glamorous job in China. A chef got as much respect as a car mechanic—they were replaceable cogs in the kitchens. Lacking workers' rights (there was no such thing as unions) and without much education, chefs rarely rose to prominent positions. Chinese restaurateurs, most of whom had experience in the business world but little knowledge of how to cook, were beginning to receive recognition and celebrity status in China, but chefs remained firmly in the kitchen and outside the limelight. The goals of my classmates were practical rather than lofty: food was a long path to a hoped-for life in an apartment in a high-rise building with an elevator—contemporary China's equivalent of a suburban house with a white picket fence.

As it happened, I was signing up for classes in Beijing just as professional cooking was making a comeback in mainland China. Chinese-American friends snobbishly advised me to go to Hong Kong or Taiwan if I "really wanted to learn how to cook proper Chinese food." My family disapproved of my venture. My father, a physicist, had sent me to an Ivy League school— in no way did he want to see me become a chef, in his eyes the lowliest of Chinese occupations. My maternal grandmother, who briefly owned a restaurant in southern China in the 1940s, admonished me. "You can never trust a chef!" she said, shaking her head. My grandparents had chosen a poor time to enter the restaurant business: Chairman Mao had been making headway into the south with his guerrilla army, and the family was eventually forced to flee to Taiwan. But my grandmother didn't blame the upheaval for the restaurant's demise; she claimed that it was because the chef stole tins of abalone from the pantry.

My family thought that the Communists had ruined food in China, despite the fact that few of them had spent time in China since the Communists took power. They were right to a certain extent: owing to socialist policies, mainland chefs had little passion for their jobs, had limited access to quality ingredients, and cooked for the masses. But mainland China was still the birthplace of a nebulous cuisine that had been bastardized, refined, reinvented, and sanitized around the globe. Perhaps the cooking methods I would learn in Beijing weren't as sophisticated as in the fancy restaurants I went to with my grandparents in Taipei, but the meals I had eaten in mainland

China's back alleys were just as satisfying. And Beijing—with its Maoist legacy and the chaotic hurry to modernize that made it a hub for Chinese from every nook and cranny of the nation—was certainly the most interesting place to learn to cook. I was curious how the past half century of turmoil and present economic development had affected the food.

When I began writing about food, two years before I enrolled in cooking school, I started by interviewing old Chinese chefs, most of whom began their careers not long after the Communist Revolution. I'd ask them how they got interested in cooking. I quickly learned that it was a stupid question. The answer was always the same: "I wasn't interested. I didn't have a choice." Under China's state-planned economy, the government drafted people for their jobs, as if the nation were a giant, compulsory army fighting for the greater good of socialism. I met chefs who started their jobs without knowing how to hold a knife. Not surprisingly, many of China's best chefs fled to Hong Kong or Taipei around the time the Communists came to power in 1949.

But things have changed in recent decades. Farmers, forced into communes under Mao, were again allowed to grow and sell their own vegetables and raise their own cattle starting in the late 1970s. Young, energetic chefs who had freely chosen their vocation were replacing old, phlegmatic cooks who'd been commanded to stir-fry. Food rationing ended in the 1980s, and meat, once a luxury, became a regular part of Chinese diets. Around the same time, restaurants, previously all owned by the government, were being passed to individual owners, and eager entrepreneurs opened their own establishments, catering to urban families that were finally able to afford meals out. (Some state-owned restaurants still existed, and could usually be identified by their ostentatious entryways, musky smell, and bad fluorescent lighting.) With the end of the state-planned system came the end of on-the-job training. Hundreds of cooking schools like Hualian opened to teach the chefs needed for private ventures.

By the time I enrolled at Hualian, the industry was oversaturated with restaurants and chefs. Within a half mile radius of my home, I could count more than fifty restaurants. Nearly every street in Beijing was a restaurant row, with tiny takeout windows selling *xiaochi* ("little eats") jammed between banquet-style restaurants with fancy lobbies and more modest, unadorned family ventures, punctuated by the occasional convenience or clothing store. Eating out was so common that my Chinese friends lamented that the tradition of inviting guests over for a home-cooked meal was slowly disappearing. Even before the decline into darkened Communist times, eating out had not been common; most Chinese had lived in rural areas, and if they ate out at all, it was at simple stalls. Most banquet-style restaurants, located in the cities, were limited to a small, elite class. But in booming post-Mao China, the traditional Chinese New Year's feasts that had always been eaten at home were now eaten at restaurants. Weddings, which in the past had often been celebrated modestly in country villages, were also taking place in restaurants.

The once familiar greeting "Have you eaten yet?" was falling out of fashion, especially in big cities. Everyone had eaten already, and they were eating well.

Teacher Zhang and I had an uneasy relationship. Most of the time, he spoke with a guttural Beijing accent. But when he turned to me to ask a question, he enunciated very carefully. "Miss Lin," he'd say with a hint of condescension, as if he were taunting me. He'd pause and take a sip from his glass jar filled with tea or wipe his hands on the sleeves of his ski jacket. "How is *food* different from *cuisine*?" Occasionally he'd look at me with his beady eyes and let out a little laugh, shaking his head. The teachers and students were baffled as to why I called myself "Chinese American," a fuzzy concept in their heads. They seemed unable to conceive that it meant that I knew English better than I knew Chinese, much less that I could be more American than I was Chinese.

My Mandarin was not bad, but it was far from perfect. I could hold fluent conversations, even if my tones were a little off. But I had neglected to work on my reading and writing skills after my first year of living in China, and nothing in my previous experiences had prepared me for the nuances of discussing fish guts. While my classmates dutifully copied down what was written on the board, my pen often hovered above my notebook, midcharacter, I had trouble finishing basic culinary words like "sauce" and "steam."

A month into the class, after about the fiftieth time Teacher Zhang had turned to me to ask, "Do you understand?" and received a blank stare in return, something seemed to dawn on him.

"Miss Lin, Chinese is not your mother tongue, is it?"

The revelation rocked the class, setting students atwitter. Never mind that I had clearly informed the administration of my identity and my purpose when I enrolled, and that the information had been funneled and disseminated in the usual bureaucratic Chinese way. "Miss Lin is a Chinese-American writer, and she wants to spread propaganda about Chinese food to the American people," an administrator had proudly announced to the class on my first day.

I had needed assistance to fill out the registration forms. I had assumed that when I interrupted Teacher Zhang to ask questions, he and the students understood that I had to process the information in Chinese first and then mentally translate it into English. Apparently, however, they had simply thought I was retarded.

After his shattering discovery, Teacher Zhang called for a break. He threw his legs around a chair, sitting with his back to the chalkboard, and studied me carefully. One student wiped off the board. Another student offered him a cigarette. The classroom filled with smoke as half the class lit up.

"So you've spent a lot of time in America, have you?" Teacher Zhang asked.

"I was born and raised there," I replied.

He narrowed his eyes. "But why do you look Chinese?"

"My parents are originally from China."

"Why isn't Chinese your mother tongue then?"

I was born and raised in America, I repeated.

One of the students, Little Pan, perked up. His name described his youth more than his girth, which was enhanced with generous helpings of Chef Gao's samples. "If an American man has a baby with a Chinese woman, what does it look like?"

Teacher Zhang pushed on before I could figure out how to respond. "Where are your parents from?"

"Guangdong and Fujian, or at least their ancestors were," I replied. My parents had grown up in Taiwan and moved to the States in their early twenties, but I didn't want to shock the classroom any further. Taiwan was a touchy subject, since China considered it a "renegade province" that remained part of the "motherland" and had successfully indoctrinated 1.3 billion people in the intractable belief that this was indeed the case. I usually felt it was wiser to skip the subject of Taiwan altogether, to spare myself the inevitable tirade that arose from a casual mention of the island.

Teacher Zhang and the students dragged on their cigarettes, as ashes fell to the floor. They stared at me, perplexed. America meant white, the land of people who looked like President George W. Bush. China meant Chinese, and I looked Chinese. My explanation didn't seem to sway them; they eyed me with suspicion. Why was Miss Lin pretending to be something she was not?

I needed something to make it concrete to them. I happened to have my passport with me, but I hesitated to pull it out. An American passport meant status to the Chinese. It meant being a member of the most powerful country in the world. I was uncomfortable with the idea that some Americans thought their passports provided them with immunity when they traveled abroad, especially to developing countries. I knew that flaunting my passport in China was the equivalent of boasting that I was landed gentry in a room full of Victorian factory workers.

In desperation, I handed the blue booklet to Teacher Zhang. The students gathered around, admiring the stamps as he flipped through it. "Wow! She's been to Thailand!" someone exclaimed. My classmates looked at me with new respect. In a matter of minutes, I had gone from class dunce to passport-wielding, bona fide American.

Somehow this only fanned the flames of my exasperation after enduring weeks of Teacher Zhang's condescension. "I've had a really hard time here," I said. "This may be an easy class for all of you, but try taking it in another language. I did my

schooling in the United States, and none of it was this hard."
And before I could stop myself, I blurted out, "Even graduate
school wasn't this difficult."

If this was an insensitive thing to say in a class of Chinese
workers who would probably spend months doing manual
labor in the kitchen after graduating from cooking school, who
would never have the chance to go to college, no one let on.
They looked smug, as if it reaffirmed something they knew all
along: of course cooking school was harder than American
graduate school!

"I thought Miss Lin was pretty from the moment she
stepped into class," said Tie Gang, a short-order cook at the
Ministry of Railroads with a buzz cut and a beer belly. He had
already made it apparent that he had a crush on me by waiting for me after class every day and following me wherever I
went. Though I was not so impressed with Tie Gang, my classmates were: he had become a leader among them when he announced that he was already a cook and was there to sit for the
national *advanced-level* cooking exam.

My cheeks flushed. "What about all of you?" I asked, direct-
ing my question to the rest of the students. "Where are you
from?"

"Beijing," said the guy seated behind me who showed up for
class in fatigues. He was a soldier in the People's Liberation
Army.

"Beijing," said a short guy with fluffy hair streaked with red
highlights.

"Beijing," said Little Pan. He worked in the maintenance
department of an office building.

"Dongbei," replied a tall, skinny guy who worked in security at an upscale mall. At least one other student was from

somewhere else, even if it was just an overnight train ride away to the northeast.

"But your roots!" Teacher Zhang interjected, pronouncing "roots" with a particularly harsh pitch. "Your *roots* are still in China."

"Yes, that's why I'm here," I said.

He smiled. "Americans don't understand Chinese history. You don't study any history but your own. And you only have three hundred years of it!"

"Do you study American history in China?" I asked.

"Yes, of course we do," he said.

"What year did America gain its independence?" I quizzed him. My classmates' eyes went wide with shock that I was challenging a teacher.

"Let's get back to the lesson," Teacher Zhang grumbled. From then on, he peppered his lessons with references to Chinese history and allusions to emperors, poets, and the Buddha. Each time, he glanced in my direction, chuckled, and said, "But Miss Lin doesn't understand . . ."

I had moved to China in 2000 as a budding young writer who had just graduated from journalism school. After a year of studying Mandarin in Beijing, I moved to Shanghai and began freelancing for American magazines and newspapers, which were demanding more stories from China as its economic might grew and America awakened to the possibility of a new, emerging superpower taking its place. It was a good place to start my career.

But I found the experience of living in Shanghai alienating. I was weary of arriving at meetings to be greeted by people demanding to know where "the American" was to whom they had spoken on the phone. I was tired of Chinese security guards detaining me at the gates of foreign diplomatic compounds, unconvinced that I was going to visit friends until I produced my passport, while more obvious foreigners breezed through. I was sick of people wondering why I spoke Mandarin with a funny accent and why I didn't understand their idioms.

Most of all, I had difficulty relating to many Chinese, who had been indoctrinated in an educational system that relied on propaganda rather than on fact, who had no foreign perspective, and who earned a tenth of what I made as a freelance journalist. I did make several close Chinese friends, but they struck me as atypical, and it didn't surprise me when they left China to study or work in the United States or elsewhere.

I sought refuge in the expatriate community. But I didn't look like the usual *laowai*, the Chinese term for foreigner, and I didn't have the lifestyle of one. There were plenty of Americans, Europeans, and Australians cavorting around Shanghai on lavish expense accounts, their salaries padded with "hardship pay" for having to live in a developing country. They lived in luxury apartments or villas, dined in foreign restaurants, and rarely vacationed anywhere in China (or, as they called it, "the provinces"), opting for a flight out to Bali or Thailand. Many did not speak Mandarin, and they often joked about the funny habits of the Chinese. I straddled the expatriate bubble and the Chinese world outside it, not quite belonging to either. So it was in China, ironically, that for the first time I felt the urge to call myself a Chinese American. It was the first time I had to seriously grapple with issues of race, identity, and where I fit in.

. . .

It was the alienation I felt that led to my rabid obsession with Chinese food. I imagine my subconscious thinking went something like this: if I can't connect with the people, at least I'm going to connect with the food. I hadn't been a foodie before I moved to China. But in my desire to identify with something Chinese, I took up the cuisine with a fervor that came second only to my passion for writing.

I had eaten a good amount of Chinese food growing up in Southern California. My mother, a biologist and later a software engineer, didn't have much time to cook, but she had a few dishes in her repertoire that she made again and again. One was a pork patty that she steamed in a shallow dish that fit in a rice cooker. When both were done, she poured the pork juices over the rice. She also made simple stir-fries with oyster sauce, and fried rice with ham, eggs, peas, and carrots. On special occasions, my mother would bring out an electric pot in which we would cook thinly sliced meats and leafy vegetables at the dining table, in a Chinese version of fondue called hot pot. Once in a while, my whole family—including my maternal grandmother, who came for extended visits—would spend a Saturday or Sunday wrapping dumplings with minced pork and shiitake mushrooms. Boiled or pan-fried, they were as comforting as mashed potatoes.

Though I had been accustomed to Chinese food, I didn't appreciate it as a child. I was embarrassed by the smells that filled our house, and worried that they would drive my non-Chinese school friends away when they came over. We ate home-cooked Chinese meals about half of the time, and the rest of our meals consisted of pizza (delivered), the occasional Western dish my mother ventured to cook (macaroni and cheese, spaghetti and meatballs, meatloaf), and trips to restaurants. When we went out to eat, it boiled down to two options: Anthony's Fish Grotto or a Chinese restaurant. When my parents missed the food of their childhoods, they would pile the family into the Toyota station wagon and drive a hundred miles to Monterey Park, a Chinese enclave near Los Angeles. I hated those trips. The car ride was far too long, and my father would sometimes lose his temper as he steered us into the parking lot of a lack-luster strip mall, to discover, after a two-hour drive, that it was impossible to find a space. The restaurants were loud, the food came slowly, and I went through a phase when I didn't like fish, which was shown to us live, flapping in a bag, before it was steamed. I wanted McDonald's or Taco Bell. Only later, after I had left home and was attending college in New York City, did I embrace my cultural roots. By the time I reached college, I was no longer ashamed of the smells, appreciated the delicate way fish was steamed, and loved Sunday dim sum brunches. I frequently took the number 1 train to Canal Street, to my fa-vorite restaurants in Chinatown.

MOM'S STEAMED PORK PATTY

- ¾ pound lean ground pork
- 2 tablespoons water
- 1 tablespoon rice wine or sherry
- 1 tablespoon soy sauce
- 1 tablespoon cornstarch
- 1 teaspoon sugar
- 1 teaspoon rice vinegar
- ¼ teaspoon salt
- 3 cups uncooked rice

You'll need a rice cooker with a steamer basket tray and a steamer-safe nonstick bowl for this dish.

In a large bowl, mix together all the ingredients except the rice, then transfer the mixture to the steamer-safe nonstick bowl. Put the rice in the rice cooker and prepare to steam it according to the manufacturer's instructions. Place the bowl with the pork mixture on the steamer basket tray and start the rice cooker. When the rice is ready, the pork will be cooked. Serve immediately, pouring the juices of the pork over the rice.

◎

Even with all my experiences eating Chinese food, I was not prepared to eat in China. It had taken me close to two decades to feel comfortable with Chinese food in America; now, in China, I faced a whole new set of challenges. In the beginning, I had felt as disconnected from the food as I had from the people — my taste buds were at first overwhelmed by flavors that felt too chaotic, too intense. Ordering at restaurants was a minefield; menus were full of items with beautiful, ornate names but arrived in the form of innards, claws, and tongues. I once settled on what sounded like a safe appetizer, drunken shrimp, without realizing that the name was literal. A few minutes later, a waiter brought a covered glass bowl, shook it, and placed it on the table. A couple dozen shrimp, dazed and drenched in rice wine, attempted to crawl up the sides of the container. The first time I sampled an authentic kung pao chicken, my teeth clamped down on a Sichuan peppercorn, instantly numbing my mouth, which made me associate Sichuanese restaurants with the dentist. Even my beloved dumplings were different: they were served with vinegar and heaps of minced garlic — no soy sauce — and contained pungent vegetables like Chinese chives. What was this alien food that was supposedly authentic?

Nevertheless, I began to adapt to the textures and flavors that I had initially found bizarre. By the time I started cooking lessons in Beijing, I no longer went more than a day or two without sprinkling chili peppers on my food. I had come to enjoy the tingling sensation induced by encounters with the Sichuan peppercorn, which I later equated with the rush from a double espresso. I discovered there was a name for this sensation—*ma*—and that it was considered one of China's seven basic tastes. Dumplings came in a range of fillings that were better than the ones that I had eaten growing up. I overcame my irrational (and possibly self-loathing) fear of ginger, which used to nauseate me, even in the smallest amounts.

I began devouring with delight the banquet dishes that had made me cringe in the beginning—dishes like jellyfish heads with vinegar, braised sea cucumbers, steamed chicken feet, and fermented bean curd (aptly called stinky tofu in Chinese). I fell in love with the meaty, flavorful Chinese eggplant, whose American cousin did not charm me in the same way. I began to lust after the pomelo, a sweeter relative of the grapefruit, and the loquat, an orange-colored, oval-shaped fruit with the fleshy texture of a plum and the tangy flavor of citrus.

Though I was developing a passion for Chinese food, I was growing weary of Shanghai after three years. I decided to move to Beijing, where I had studied for a summer during college and the year after graduate school. It was a good city for learning. China's best universities were located there, the back alleys had a laid-back feel, and the expatriate bubble was smaller. Beijing, where I had learned Mandarin, would also be the place where I learned how to cook.

As it turned out, cooking school was the first place where I truly felt immersed in Chinese life. Despite the cultural barriers and frustrations, I found it strangely invigorating: copying

Teacher Zhang's nonsensical babbling, gaping at Chef Gao's cooking, scrambling for free samples. My Mandarin was improving, and I was introduced to a whole new vocabulary. I had finally become a participant in this baffling, contradictory country that was transforming itself into a superpower.

I was even grasping some of the basic principles of Chinese cooking, but I continued to be impatient for hands-on experience, unlike my classmates. They seemed conditioned to China's absurd teaching methods, and I later learned that most of them had enrolled solely to pick up hints on how to pass the written part of the certification test. We'd been promised that after six weeks of instruction, we would have two classes devoted to improving our knife and stir-frying skills. I didn't want to wait that long. After being rebuffed by various teachers at the school, who clearly considered it a waste of time to hang out with a frivolous foreigner who wanted to learn how to make kung pao chicken, I decided to seek the advice of Chairman Wang one afternoon when class was dismissed. "Chairman" was a bit misleading; it was more of an honorary title for a low-paying, all-purpose job that encompassed serving as registrar, assistant to the school's president, assistant teacher, food purveyor, and de facto janitor—in short, all the tasks that no one else wanted to do. During demonstration classes, Chairman Wang moved around the kitchen in a slow shuffle, tidying up after Chef Gao and lighting the burner just when he needed it. She had a stern, matronly air about her, but once in a while she'd break into howling laughter. She always wore a blue lab coat, which, combined with eyeglasses and wiry gray hair that stood up in stiff, Albert Einstein–like puffs, made her look like a mad scientist.

"You want cooking lessons?" Chairman Wang asked, as if this were a preposterous request at a cooking school. She continued to mop the grimy kitchen floor, which seemed to retain the same amount of dirt no matter how many times it was cleaned. I couldn't tell if she was taking my request seriously. For that matter, I wasn't sure she—or anyone at the school—took me seriously, being not only a foreigner but a woman to boot.

It surprised me that the idea of a woman in the professional kitchen was such a taboo. After all, even critics of Mao conceded that he had advanced women's rights: he eliminated the tradition of foot binding, banned prostitution, and gave women equal access to education and jobs. During the Cultural Revolution they had been forced to toil equally in the fields. Mao's changes had a lasting effect; rarely did I meet a woman who didn't have a job, and female doctors and other professionals were common. But I was learning that gender equality didn't apply in the kitchen.

"You want to be a chef?" Teacher Zhang had asked me once.

Did he think it was possible? I asked.

"You could make pastries," he'd replied dryly. Given the poor quality of northern Chinese pastries, that was like saying I could be a burger flipper at McDonald's.

"You could work in a Western restaurant," a classmate had suggested. "Women aren't cut out to be stir-fry masters."

But didn't women cook at home?

"Yes, but the fire is much smaller," the classmate had pointed out. "It's a tough job being a chef."

Chairman Wang paused in her mopping, staring at me through her thick glasses.

"All right," she said. "I'll teach you."

2

FOR OUR FIRST LESSON, Chairman Wang had let me choose the dish. I bought the ingredients for deep-fried shrimp. But we hadn't settled on a fee, and as I walked across the school's basketball courts to the kitchen, I tried to figure out the best way to bring up the issue of money.

"I'm really grateful that you're spending time with me, and I'd like to pay you," I said.

Chairman Wang pursed her lips and remained silent. What was going on behind those thick glasses? Surely she wasn't planning to teach me for free, was she?

Money was an odd topic in China. Usually the Chinese inquired relentlessly about what things cost and how much people made, even among strangers. Shoppers haggled over the price of groceries, clothes, and bicycles. But at moments like this, talking about money seemed taboo.

Three of my classmates caught up to us as we entered the kitchen. Our theory class had just adjourned, and the students were intrigued to hear that I was having a private cooking lesson. I hadn't let on that I was taking a "private" lesson, but the students had found out from the gossipy teachers. I was embarrassed, because even in post-Mao China I thought it sounded too bourgeois, but my classmates had an unexpected reaction: "We want to learn too," said Little Pan. But I knew they wouldn't be willing to pay anyone for individual lessons when they had already spent a good chunk of their money on their courses. My embarrassment was replaced by guilt; I could afford the lessons and they couldn't. But I also felt possessive of Chairman Wang.

"I arranged this class especially with her," I whispered as the chairman disappeared into the backroom to get an apron. "I'm paying for it, you know."

"How much?" they asked loudly, practically in unison.

"We haven't discussed it yet," I said.

How much had I paid for my butcher knife? they wanted to know. (Four dollars, I said.) The chairman reappeared and asked me the same question. Then she asked me how much I had paid for the shrimp. (A dollar-fifty.) Always in these in-quiries, my interrogators admonished me for paying too much. Being a foreigner, I was bad at bargaining, invariably the fool. (Never mind the disparity in our incomes. The money I made from writing articles for American publications put me in China's top income bracket. I soon learned that it was a bad idea to tell people how much I made or how much rent I paid, even if it wasn't a lot by American standards.)

Chairman Wang took one look at my cleaver and rushed into her office, her blue lab coat fluttering behind her, and re-turned with a replacement. "Yours is pretty good. But you won't be able to use it today. It's not sharpened."

"Can't we sharpen it here?" I asked.

The chairman explained that professional knives came out of the factory with blunt edges. The ordinary stone block that the chefs at the cooking school used to hone their cleavers wasn't strong enough to put an edge on a never-sharpened knife. "You'll have to find a professional sharpener and tell him that the mouth of the knife needs to be opened."

As the chairman instructed me on the cooking, my classmates draped their arms around each other and watched. It still threw me a bit, the display of affection between men in China, which would have been construed quite differently in America.

"Look at the way your shrimp are moving!" Tie Gang, the class leader, scolded. He detached himself from the group and grabbed the knife out of my hand. "You must keep your left hand firm while cutting with your right hand."

Eventually the three of them got tired of watching the lesson and left. Chairman Wang continued to correct me, on everything from how to hold my knife (like a giant razor, between my thumb and index finger) to my posture. My culinary skills had been limited to making basics like pastas and stir-fries, and baking cookies and brownies—out of a box. When I was growing up, cooking had never been emphasized at home, since I was supposed to become a doctor or a lawyer, and my years in China hadn't improved my culinary skills, since eating out was so cheap and easy.

I had watched Chef Gao make deep-fried shrimp and had chosen it because it seemed simple and familiar, but I couldn't do anything right. I tried to stand like Chairman Wang, putting my right foot parallel to the table and my left facing 90 degrees outward, like a poised ballerina.

"Keep your belly away from the chopping block," the chairman commanded. I sucked in like an awkward girl at a school dance.

Once the shrimp were breaded, Chairman Wang lit the pilot light. I tried to raise the wok with one hand, but it didn't budge.

"Don't use the handles. You won't have enough support," she instructed.

I tried again, lifting the side of the wok with a folded kitchen cloth. After another series of admonishments, she pushed me aside and took control of the wok and spatula. I stood on the sidelines, occasionally dropping a battered shrimp into the wok. This wasn't a cooking lesson; this was what six-year-olds did with their mothers.

The shrimp came out crisp and nicely browned, though they were a bit tough. I hadn't given the poor suckers a thorough enough pounding with the back side of the cleaver. "*Hai xing*. Not bad," the chairman commented, as if I had done all the work. "Cooking is like driving a car. You just have to learn the mechanics. It's that simple."

"Isn't it a talent, an art?" I asked.

She raised an eyebrow at my naïveté. Okay, then, did she know how to drive?

"No," she said, shrugging. Then she added abruptly, "I'm not going to lie to you. This class is going to cost money."

"Sure," I said, holding my breath for a second. "Name your price."

"Well, a lot of teachers would charge more than me. I'm only asking for thirty *yuan*."

Less than $4 for two hours of private class; I was stunned. I happily handed her the money, and, the issue resolved, we both relaxed. We chatted as we cleaned up.

"Are you married?" she asked.

Like money, this was another topic that Chinese people talked about bluntly, and I still couldn't get used to it, especially when I knew that they looked at me with pity when they found out that I was twenty-eight and single. I had actually started subtracting two years from my age sometimes, remembering that at twenty-six my unmarried state hadn't caused as much of a fuss.

But Chairman Wang knew how old I was, and all she said in response was "That's okay. I didn't get married until I was thirty-three. My husband is six years older than me."

We smiled at each other. It was consolation from a former spinster to a younger spinster: I had plenty of time.

DEEP-FRIED SHRIMP

12	jumbo shrimp, shelled and deveined
½	teaspoon ground white pepper
½	teaspoon salt
2	teaspoons rice wine or sherry
2	scallions, white parts only, finely shredded
1	teaspoon ginger, peeled and minced
1	large egg and 1 egg yolk
1	cup all-purpose flour
2½	tablespoons cornstarch
1	quart vegetable oil

Butterfly the shrimp so they can be spread out flat, and tenderize them by pounding them lightly with the flat side of a cleaver. Arrange the shrimp on a plate and rub with the pepper and ¼ teaspoon of salt. Sprinkle with the wine, scallions, and ginger, then cover with plastic wrap and place in the refrigerator for 30 minutes, or up to 2 hours.

In a bowl, beat the egg and egg yolk. Add the rest of the salt. Stir in ½ cup of the flour and the cornstarch; mix thoroughly. The batter should be a little thinner than pancake batter. If it is too thin, add more flour and cornstarch (three parts flour to one part starch). Pour the remaining ½ cup flour onto a plate. Place the batter bowl, flour plate, and marinated shrimp next to the stove.

Place the oil in a wok over medium-high heat until it is hot but not smoking. As the oil heats, dredge the shrimp one at a time in the flour and set them on the edge of the plate. When the oil is hot, dip each shrimp into the batter, holding it by the tail. Wipe off any excess batter with your fingers and gently slip the shrimp into the wok, being careful not to splash the oil. Once the wok is filled with shrimp (depending on the size of your wok, you may have to fry in two batches), increase the heat to high and fry until golden brown, about 5 minutes. Drain the shrimp on a paper towel and serve immediately.

Cooking school introduced me to many nuances of Beijing life I had missed when I was spending my time with other foreigners. I began commuting by bicycle, as an ordinary Beijinger would have. Though cars were gaining popularity in China, most people couldn't afford such a luxury. I bought my Shanghai Forever bicycle from a local repairman who sat outside my apartment building fixing everything from radios to office swivel chairs. The bicycle was probably stolen. At $10, I didn't ask. The curved handlebars allowed me to sit up straight, and my groceries fit snugly in the basket. In flat Beijing, I had no need for gears, and surprisingly, no one stole the bicycle when I forgot to lock it up. (Thieves had begun preying on fancier models stored in bicycle sheds.) Even after a friend gave me a new eighteen-speed bike that cost ten times as much, I continued to ride the old one.

My apartment occupied the top two floors of a building less than two miles from the school. There was no elevator. I kept my bicycle in the courtyard, where my neighbors grew tomatoes and peppers, and climbed the six flights of stairs, often carrying two bags of groceries. My neighborhood contained several rows of new, grayish-blue apartment buildings of middle-class families. Just a few years before, it had been a maze of narrow alleys —called *hutong*—that divided traditional single-level courtyard dwellings. But in the name of urban renewal, this particular section of alleyways, which were rundown and lacked modern conveniences like private toilets, had been razed, and the government had built these low-rise buildings in their place.

To the west, where my cooking school was located, some traditional neighborhoods—and the human-scale village feel of old Beijing—still existed. To the east lay the new skyscraper-clad city. My residential compound was just within the Second Ring Road, which marked where Beijing's ancient wall once stood. After Mao came to power, the wall was demolished and the wide road was built in its place.

More transformations were taking place before my eyes. When I moved back to Beijing in 2004, the area near my apartment was full of empty lots. Now construction workers were putting the finishing touches on a row of massive skyscrapers; farther east, outside the Second Ring Road, hundreds more tall buildings were going up, including the China Central Television headquarters, a zany pair of slanted towers that were connected by an enclosed platform suspended in the sky. The empty, freshly paved avenues where the dust of new buildings was still settling felt surreal. I was living in the middle of the world's biggest construction site.

To get to the cooking school, I bicycled away from all this development into the traditional heart of the city. When I was pressed for time, I took the wide new road. When I wanted a pleasant, leisurely ride, I glided through the alleys.

Bicycling in China was different from cycling in the United States. On the boulevards, Chinese cyclers moved in herds.

Nobody yielded, but no one collided either. At intersections, more bicycles gently folded into the stream, as crossing guards blew their whistles and waved their flags in vain. Cars honked and inched along. Motorcycles revved their engines and sped past. Motorized bicycles silently edged up next to me. In the beginning, I developed bike lane rage whenever a bus or a car drifted in front of me, but I learned to weave around the vehicles like everyone else. The goal was to keep everything in a delicate balance of motion.

Fewer bicyclists pedaled through the *hutong* areas because the alleys were narrow, making it a slower ride. One afternoon as I biked down a brick-lined lane, a crowd gathered to watch a game of Chinese chess. The elderly strolled the dusty alleys. Tanned, shabbily dressed men and women on bicycles pulled junk in wagons behind them, shouting indecipherable phrases that sounded like heckling at a baseball game. But I was too focused on the path to think much about it. Though more pleasant than riding the wide roads, the alleys were, like almost everything else in China, chaotic: cyclists careened around corners, and countless potholes required constant alertness. When a friend bicycled home drunk one evening, she ended up flying into a construction ditch.

With a specific mission one afternoon, I rode down the alleys. In cooking class, we were finally going to get a practical lesson in how to use a knife. I had already learned a lot of knife "theory" in the first few weeks of class. Of all the skills on which Chinese chefs were judged, the most important was their *daogong*, or cutting technique. I knew that chefs from different parts of China used knives tailored to their specific regional cuisines. In Shanghai, a chef's knife had a pointy tip that resembled the profile of a shark's head. In Sichuan, the most common knife had a blade with the contour of a bell, and a Cantonese knife had a narrower blade with a sharp tip that resembled Western knives. In Beijing, chefs cut with rectangular blades that were so wide and clunky they reminded me of props in a horror movie.

I learned that Chinese chefs never simply sliced a vegetable or a piece of meat. Though chefs were vague when measuring ingredients, they were exacting in their cutting techniques. Boning chickens, fish, and eel required different approaches. Chefs used dozens of terms for cutting, based on the angle of the slice — horizontal, diagonal, or perpendicular — and the motion of the cut — pulling, pushing, sawing, and waving, among other movements. I once watched a chef friend demonstrate the "circle cut," inserting a cleaver between an apple and a cutting board and shearing off the skin of the fruit by rolling it over the knife.

Knives were especially important in prep areas, since they did not appear on the dining table. Everything in a Chinese meal was already cut into manageable pieces and picked up with chopsticks. "We don't eat with knives in our hands," my Taiwanese father had said once. "Because that is for barbarians." (Apparently he didn't find eating with sticks primitive.)

Yet for all I had learned about knives, there were still a couple of basic things I didn't know. First, I didn't know how to use one properly. Watching the jagged shreds of ginger, leek, and pork that fell from a blade borrowed from the school during a private lesson, Chairman Wang commented that I would be lucky if I could make sixty dollars a month cooking in a cafeteria. After watching a little longer through her thick glasses, she was more generous: "Maybe a hundred dollars, if you don't want room and board."

The other thing I didn't know was where to get one sharpened. When I asked Chairman Wang, she replied, "The sharpeners are everywhere." I assumed she meant I could find a sharpener at a roadside stand or a shop in my neighborhood, so one afternoon I wrapped my cleaver in newspaper, put it in my backpack, and set off on my bicycle into the *hutong* neighborhoods. After coasting down a number of alleys, I hadn't found a trace of a sharpener.

I saw a man squatting near the wall of a courtyard house, fixing a piece of machinery. When I asked for directions to the nearest sharpener, he looked apologetic. "I really don't know," he said.

On the way back home, I poked my head into my favorite hole-in-the-wall Sichuanese restaurant, across the street from my apartment building. The lunch shift was over, and the chefs and waitresses lounged at a table covered with dirty plates, bowls, and glasses.

I took out my knife and showed it to them. "Where," I asked, "do I get this sharpened?"

One of the chefs picked up the knife. "Not bad. Good for household use," he said, making chopping motions in the air. "How much did you pay for it?"

The boss of the family-run joint, a woman in her thirties whose voice was always hoarse from shouting orders, said, "*Ai*, we haven't seen the knife sharpener for three days. He should be coming soon." She explained that I hadn't been able to find a sharpener because they didn't have shops or stands; they biked around the city, weaving through neighborhoods, retracing the same path every few days. There were fewer of them now because most households had started buying ready-to-use knives. She told me I had to listen for the clanging sound made by the knives that the sharpener had strung together and rattled at his side while he bicycled along.

"Huh," I said. "I've never noticed it."

She gave me a sarcastic look that implied, *Of course not. They're invisible to people like you.* "You can leave your knife here. The next time he comes, I'll have him sharpen it."

"Thanks, but I need the knife today," I said.

On the way to school, I stopped by the cooking supply store where I had purchased the cleaver. Perhaps they could sharpen it.

"Sorry," said the clerk behind the counter. "*Mei banfa.* Nothing I can do."

"You mean you sell knives that people can't use?"

"You should just go out on the street and look for someone to sharpen it," he said, shrugging. Then, seeing the look of exasperation on my face, he changed his tone. "Well, I can't sharpen it for you, but I can exchange it for the ready-to-use kind."

So I turned it in for a light, pre-sharpened Taiwanese model and bicycled to class. The students were impressed when I pulled it out of my backpack, as if I'd upgraded from a clunky Cadillac to a shiny sports car. "How much did you pay for it?" somebody asked.

Finding the right equipment was only half the battle. I still couldn't chop. Unsurprisingly, Tie Gang, the student preparing for the advanced-level exam, turned out to be the deftest with a knife. He had been a persistent admirer, continuing to follow me out of class every afternoon and inviting me to his home for "private" cooking lessons. Watching his skills, I almost regretted declining his invitation. He cut his carrots in a quick, rhythmic fashion — *chop chop chop,* with the consistency of a metronome. My chops were syncopated, but not with any particular beat: *chop, ch-chop, chop-chop*. As I hacked away, he sidled up to me and laid a ghost-thin shred of carrot on my block. He gave me a knowing look and went back to his station.

After a month or two of chopping, I felt more comfortable with the cleaver than with the narrower knives I once used. The cleaver felt safer, since I could put the back of my fingers against the surface of the knife, the weight of the blade making it steadier and smoother than a Western knife. Once I was done chopping, I pushed the ingredients onto the cleaver's flat surface and seamlessly transferred them into the wok.

I was practicing my knife skills in my kitchen one afternoon when I heard a strange clanging echo six floors below my window. It sounded like a dozen tin cans banging against one another. I looked out and saw a man on his bicycle, his knives flapping at his side as he pedaled down the newly paved avenue. For a minute, I wished I had kept my blunt knife for him to sharpen, though he would have been gone by the time I made it to the bottom of my stairs. As he drifted down the street on his rusty bicycle, trucks and shiny sedans overtook him. I didn't see anyone stop him for his services.

As it turned out, Chairman Wang was an excellent teacher once she began to let me do things on my own. She never seemed to get bored, even after she had gone over the same skill dozens of times. She evaluated my cooking honestly without being too harsh. If a dish turned out to be edible, she would say "*hai xing*" — not bad. And she peppered our lessons with interesting stories from her past. Her voice was always soft and pleasant, with her folksy Beijing accent adding an occasional rounded r sound to the end of her words.

"Did you ever cook in a restaurant?" I asked her one afternoon. I was slicing a pork tenderloin for the popular Sichuanese dish *yuxiang rousi,* fish-fragrant pork shreds. I pushed the pork against the board with my left hand, held the knife in my right, flat against the board, and sheared the pork into slices as thin as sandwich meat. Then I pushed my blade through the slices to shred them.

"No, we weren't allowed to choose our jobs. We did what the Communist Party told us to do our whole lives." Wang had begun working at the cooking school twelve years before, after she had been laid off from her job at the Beijing Number 1 Candy Factory. She had started at the factory when it made bicycles and had been in charge of running the cafeteria, where the workers ate all their meals. One day, the authorities issued a decree: the factory would no longer make bicycles; instead, the workers were going to make candy. A few years later, the authorities made another announcement: the factory would be closed. Wang was then approaching fifty, which was fortunate; given China's large workforce, female factory workers could retire at that age, and her pension kicked in. Her husband, an elementary school teacher, had also retired (male schoolteachers usually retired at sixty), and the Wangs' combined pensions were enough for them to live on. But having worked hard her entire life, Chairman Wang was bored with her days suddenly free. It so happened that a friend of a friend ran a cooking school and needed an assistant, so the chairman went there to work, for a meager salary that was less than her pension.

The next lesson, we talked about cabbage. As it got colder in the fall, vegetables became scarcer—until one day every November, when huge stacks of cabbages appeared on Beijing's street corners as farmers from the nearby countryside brought in their harvest. Selling for a few pennies per *jin* (half a kilo, or a little over a pound), they were a heavy, oval-shaped variety that resembled napa. Their white stems turned a dark jade at their leafy centers. Many people would buy enough cabbage to last the entire winter, piling them on the back of their bicycles to bring them home. They were stacked outside, on window ledges and beside front doors, where they would keep for months in the freezing northern climate. When it was time to cook, all a Beijinger had to do was reach outside for a few leaves. They could be stir-fried with shrimp or pork, wrapped into dumplings, or soaked in brine to make a Chinese version of sauerkraut.

I asked Wang if she planned to store cabbage for the winter.

"No," she said. "We used to, until 1983. That's when we moved from a courtyard house into an apartment, and now there's nowhere to put it. If you put the cabbage inside the apartment, where it's too hot, it will rot. Plus, there's no need to buy all that cabbage since I mostly cook for my husband and myself. My son eats in his cafeteria at work. It would take us a month to eat just one of those giants. And other vegetables are cheaper now, so we don't have to eat cabbage all the time. When string beans are in season, you can get them for twenty cents per *jin*."

One day when we were making *basi pingguo*—candied apples —she said, "Want to hear a funny story about apples? My sister-in-law had four kids. She had them during the Cultural Revolution, when Mao encouraged families to have more children. There wasn't much to do but have lots of kids. For some reason, she liked the second child and the fourth child the most. The fourth kid got sick one day. She went out to buy an apple. People were so poor then that you could afford to buy only one apple at a time. She came home and cut the apple in half. The sick kid got the bigger half. She gave the smaller half to the second kid. The skin went to number one. And number three got nothing! Now the third kid is the only sibling making money—he went down south and prospered. It's funny, he is the only one who sends money back to his family these days."

She was peeling an apple with a big knife, the skin coming off in one long, curly ribbon. The story didn't sound very funny to me; it sounded ironic, bitter, and sad. Even so, I always looked forward to the next lesson, to hear more.

FISH-FRAGRANT PORK SHREDS
(*YUXIANG ROUSI*)

⅔	pound pork tenderloin, thinly sliced crosswise and cut into ⅛-inch shreds
½	teaspoon salt
¼	cup soy sauce
1½	tablespoons rice wine or sherry
2	teaspoons cornstarch
¼	cup plus 2 teaspoons vegetable oil
¼	cup warm water
2	tablespoons Chinese black vinegar (available in Chinese markets)
1	tablespoon sugar
1	tablespoon minced leek or scallion
1	teaspoon minced ginger
1	garlic clove, minced
2	tablespoons chicken stock
8	dried chili peppers, minced
3	dried wood ear mushrooms, soaked in water for 30 minutes, drained, and chopped into bite-sized pieces
½	cup diced bamboo shoot

In a bowl, combine the pork with ¼ teaspoon salt, 2 table-spoons soy sauce, 1 tablespoon rice wine, 1 teaspoon cornstarch, and 2 teaspoons of the oil, and set aside.

In a small bowl, dissolve the remaining 1 teaspoon corn-starch in the warm water. Add the remaining ¼ teaspoon salt, 2 tablespoons soy sauce, and ½ tablespoon rice wine, plus the vinegar, sugar, leek, ginger, garlic, and chicken stock. Set this mixture next to the stove.

Place the remaining ¼ cup oil in a wok and set over high heat. When the wok is hot, add the chili peppers and let them infuse the oil for 1 minute. Add the pork shreds and stir vig-orously for 2 to 3 minutes, making sure the meat doesn't stick to the wok. Add the wood ears and bamboo shoots and stir-fry for 1 minute, then add the sauce mixture and simmer for 2 to 3 minutes, stirring. Remove from the heat and serve promptly.

CANDIED APPLES (*BASI PINGGUO*)

  1   cup all-purpose flour
  1   teaspoon baking powder
  2   cups water
  1   large Fuji apple, peeled and cut into ¾-inch cubes
  1   quart vegetable oil for deep-frying plus ¼ cup for
       the caramel
1½   cups sugar

In a bowl, mix together ½ cup of the flour, the baking powder, and ¼ cup of the water to make a batter. Place the remaining ½ cup flour in a separate bowl. Dip the apple cubes into the flour, turning to coat them, then place them in the batter.

Heat the oil in a wok. The oil will be hot enough when a speck of batter dropped in it sizzles immediately. Gently drop the battered apple cubes, one at a time, into the oil. Fry them until they are a light golden color, about 3 minutes; they should puff up while frying. Remove the cubes and drain on paper towels, allowing them to cool. Strain the oil to remove any bits of burned batter, then reheat the oil. Fry the apples again for 1 minute, until they are nicely browned. Transfer to clean paper towels to drain.

Place a clean wok over medium-high heat and add the ¼ cup oil, swirling it around to make sure that the bottom of the wok is coated. Add the sugar and ½ cup of water and cook until the mixture bubbles, stirring occasionally. Continue to stir as the bubbles become smaller and the mixture thickens, adding a dash of oil to make sure it doesn't stick. When the mixture stops bubbling and takes on a yellow sheen, add the apple cubes and stir vigorously to coat them. Remove wok from the heat and pour the coated apples onto a well-greased plate.

Serve immediately, placing a bowl of water next to the apples. Each cube should be dunked in water before eating, to harden the caramel.

◎

My favorite new ritual was going to the nearby wet market, which was at one end of my apartment block. To get there, I walked through a small park where seniors played with their grandchildren and watched their pudgy-faced Pekingese trot unleashed. Pushing past the plastic, meat locker–like strips of the market's front door and into the warehouse, I was never sure exactly what I'd find.

I discovered the wet market just as the rest of Beijing was being introduced to Western-style supermarkets. The government planned to shut down many of the city's open outdoor markets within the Third Ring Road by 2008, in time for the Olympics. They'd be moved indoors, the way my local one had been, or replaced by sparkling new supermarkets like Wal-

Mart and Carrefour, a French chain that had already begun to colonize the city. The supermarkets had certain advantages—wide aisles, big frozen-food sections, novel goods like Italian olive oil and red wine—but I found the food at the wet markets fresher and more flavorful. Most traditional Chinese insisted on cooking with the freshest ingredients, so they shopped daily.

I went as much for the spectacle as the sustenance. Elderly women nudged me aside as they reached for long, skinny eggplants and bumpy green bitter melons. A pair of fruit vendors often squabbled from their opposite stands, accusing one another of undercutting, cursing each other's mother and father's mother. In the afternoons, once the stampede of shoppers subsided, some vendors gambled, setting up makeshift casino tables on their scales. But usually the vendors, behind huge stacks of produce and counters full of meat, were busy calling out the prices for a *jin*. "String beans, string beans, twenty-five cents!" "Tomatoes, six cents!" I was shocked by the low prices, since I had previously shopped at a small market that catered to expatriates, which felt like a general store in the Wild West, with its expensive and limited selection of American breakfast cereals, cheeses, and meats. By contrast, at this local market organic eggs sold for 75 to 90 cents a *jin*; shrimp, one of the most expensive items, hovered around $2 a *jin*; fresh shiitake mushrooms went for 40 cents a *jin*. I generally spent $2 on all the ingredients I would need to make two huge dishes with Chairman Wang.

"Do you want me to kill it?" the fishmonger's wife asked one afternoon as she grabbed a carp with her bare hands from a tank crammed with fish. She held it up by the belly for my approval, the fish flailing in her delicate fingers.

I hesitantly said yes and cringed when she slammed the carp against the floor with unexpected force for a petite, five-foot woman. She whacked it against the floor a second time, like a professional wrestler. She dropped it on a flat scale, where it flapped feebly.

"Eight *yuan.*" One dollar.

She scaled the fish with a spiky brush, giving it an occasional whack just in case it had survived the earlier trauma. She handed it to me in a black plastic bag.

"Walk slowly," she said, smiling, a Chinese way of saying "Take care."

I bought fresh chickens from a young vendor who was always engrossed in a book. One day, he, his books, and the chickens were gone. I asked the fishmonger what happened.

"Bird flu," he said. Isolated cases of avian flu had been reported around the country, and many experts worried that lax conditions at wet markets might set off a human pandemic. "The government is inspecting all the chickens. He'll be back in a few days."

"Are you worried about bird flu?" I asked.

"No," he said. "I eat fish."

As it turned out, the chicken vendor didn't come back. After that, most chickens were sold frozen in the new, slick supermarkets. The stall sat empty for months before a family of tofu makers took it over. In the mornings, they boiled and pressed soybeans through a machine to make soymilk, which was sealed and sold in plastic bags. They mixed some of the soymilk with a thickening agent and molded the curds into blocks of fresh tofu. The blocks that weren't sold were dried and seasoned, then shredded into long strands. The soymilk I bought from this stall contained no preservatives, and if I didn't drink it within two days, it would thicken into tofu in my refrigerator.

Toward the back of the market, noodle vendors kneaded large mounds of dough with their bare hands and pressed it through a machine to make fettuccine-like strands. Next door was the sesame sauce stand, which contained a large vat and a motorized arm that squashed sesame seeds into a thick, savory paste. The sauce, tossed with cold noodles, was a mouthwatering treat. Containers of sauces and bins full of spices that had once been a mystery to me were now all decipherable. Sichuan peppercorns, tiny dried berries, released a hot, numbing sensation in Sichuanese dishes; the sharp scent of star anise took the gamy flavor out of lamb and duck; vacuum-packed bags of pickled green vegetables flavored my fish soup and wok-fried string beans.

Next to the spices and sauces, the butchers hung gigantic hunks of pork, beef, and lamb—whole rib cages, an entire leg with hoof intact, round rumps. They spread out smaller cuts on wooden tables. The cuts were sold in the open air at room temperature, until one day I came into the market to find workers putting in refrigerated meat counters like the ones in the United States. But that didn't stop the butchers from handling money and meat with their bare, bloodied hands.

"You want one *jin* and fifty grams?" said my pork guy, chuckling, when I came by one afternoon. "I'm not sure if I can cut exactly that much. How about a little more? You won't mind, right?"

Sometimes it was a ploy to sell more meat—the butchers often said they weren't good at measuring and then inevitably cut more than you had asked for.

He cut a chunk of tenderloin and weighed it. "One *jin* exactly!" he said. (If I didn't believe him, I could take the pork over to an independent scale that had been set up in one corner to verify quoted prices.) He sliced off another sliver and handed the pieces to me in a plastic bag.

Down the way, another butcher greeted her customer. "You're back in town? You've gotten skinny! Your face is so skinny! That's what happens every time you come back from the village."

Maybe there would come a day when I'd crave the sparkling floors, plastic-wrapped chicken breasts, and huge aisles of frozen food in an American supermarket, but the day hadn't yet come.

When I asked Chairman Wang if she preferred the wet markets to the supermarkets, she shrugged. She was an equal-opportunity shopper: she'd go to the wet market for fresh vegetables and meats, and go wherever was cheaper for dry goods. The slick new supermarkets were usually more expensive, but they sometimes ran promotions and she'd scour the aisles for deals. If it happened that something like toilet paper went on sale, she'd buy a year's supply.

Though I loved shopping in the markets, I wasn't very good at it—at least in Chairman Wang's estimation. At the beginning of each lesson she'd examine what I had bought, asking me how much I paid, either nodding approvingly or snickering. Once in a while she would peer into a plastic bag and say, "Oh, you bought *that* kind of cabbage?" or "You should have bought *northern* tofu, not southern tofu!"

Chairman Wang offered to teach me how to shop. On a cold Friday in the middle of winter, I met her in front of her local market, held in and around a drafty warehouse where she had shopped for two decades. As it happened, the market was scheduled to be shut down for good after the weekend, to make way for yet another high-rise apartment building. A few of the stalls were already empty.

Even with doom imminent, the market hummed with life. It was larger than the market near my home and catered to individual shoppers and restaurants. In the meat section, shoppers perused aisles of pig and lamb carcasses while butchers paced around their offerings like lions lording over freshly killed prey. Vendors in black rubber boots held clipboards and counted out huge wads of cash. Restaurant buyers pushed dollies loaded with giant transparent plastic bags, ballooned with air and water, in which fish swam about. Even in the subfreezing weather, the outdoor stands bustled with buyers and sellers haggling over heaps of vegetables and fruit. It was hard to imagine that it would all be gone in a few days.

We started in the poultry section. The butcher said the chickens had been slaughtered around midnight. The whole birds, plucked but with their heads intact and their feet stiff and straight, usually sold out first, then the parts—first drumsticks, then wings, heads, necks, feet, and last of all the breasts, because they didn't have much flavor.

"You want to choose parts with no trace of feathers on it," the chairman said. "Also, examine it to make sure there aren't any bruises."

"Why is the chicken only thirty cents per *jin* today?" a customer asked. "Do you think it's infected with bird flu?"

"That's not possible," the chairman said. She had shared her views on avian flu with me in class: "I don't think we have to worry after SARS. The government will tell us the truth now." I had raised an eyebrow. When the acute respiratory illness swept Beijing in 2003, it took the government several weeks to admit that an outbreak had occurred. Even now, the government had instructed the media to deal with bird flu delicately. While local newspapers had written about outbreaks among birds in the deserts, they had not reported on the several human deaths that had been caused by bird flu.

I knew the government couldn't be trusted, but I *liked* chicken — particularly kung pao chicken, chicken wings any style, and Cantonese steamed chicken feet. I pushed thoughts of bird flu out of my head, and we each bought a few cuts of chicken before we moved on.

About half of the meat for sale at the market was pork, an ingredient that made a substantial appearance in nearly every Chinese meal. Pinkish cuts of tenderloin, marbled pork belly, and red and shiny pork kidneys were displayed on tables. The pig heads looked surprisingly peaceful, with their lips drawn back in smiles that revealed sets of straight teeth.

"Is this your daughter?" one of the vendors asked the chairman, glancing at me.

"No," she said. "She's, uh, a friend."

We perused a stall that sold beef offal. The tendons, white and hard, resembled dry loofahs. The stomach looked like brown AstroTurf. The chairman picked up several cuts with her bare hands and put them back down as if she were examining produce. No one seemed to mind.

"If you want to buy beef, come early in the morning because it sells out first," Chairman Wang said. The gamy smell of mutton lingered through the warehouse. Mutton and lamb were more plentiful in China's north, and therefore inexpensive.

We stopped in the preserved goods section to examine bamboo shoots. They were pale yellow, and their thick bases, as wide as my thigh, tapered to a tender tip. Winter was the best time to buy the shoots—by spring, they wouldn't be as tender and would have grown leaves, which had to be pulled off like the outer layers of an artichoke. The shoots, with their delicate earthy flavor, were delicious stir-fried with pork. "Pick wider, shorter ones, since we're just interested in the tips," the chairman said.

In the bean curd section, we came across a large plastic bin of tofu that had a strange reddish-brown color. "That's blood tofu," said Chairman Wang, explaining that it was bean curd mixed with pig's blood for extra flavor and texture. I learned that northern tofu was the firm kind, which retained its shape when it was diced and was good for a Sichuanese dish called *mapo* tofu. Southern tofu was the silken variety, good for deep-frying.

She carefully noted the prices of everything. "Sugar has gone up. Once the price of sugar goes up, it never comes down!" It was a cheap day for eggs. "It's twenty-five cents per *jin* today, but sometimes it's thirty." The vendor put the eggs in a plastic bag—eggs were rarely sold in cartons in China. At first, carrying eggs in a bag made me nervous, but I soon got used to it, and they rarely broke.

When we reached the fruit and vegetable section, the chairman pulled out a portable scale that consisted of a gauge and a hook. "You should never trust any of the vendors," she advised. "You see that man over there?" She narrowed her eyes toward one of the stands. "He once charged me more than he should have for tomatoes. I'll never go to him again!"

Though this would be one of her last trips to a market where she had shopped for twenty years, she didn't seem sentimental or sad. She asked each of the vendors where they planned to move.

Some shrugged. Others said they would set up at another wholesale market a few miles away.

"I guess I'll have to go there to shop too," the chairman said, sighing. "It won't be very convenient, but I'll get more exercise bicycling over there."

After two hours in the cold, my nose was running. Chairman Wang surprised me by inviting me home for lunch. "Sure," I said, glad to be going to a warmer place.

The chairman led me to a second-floor apartment in one of the drab towers that lined the Second Ring Road. The towers dated from the early 1980s, when economic reforms began, and many were now being torn down for new developments.

"Don't bother taking off your shoes," she said as we entered. She kept her own shoes on as well. It was warm but not exactly cozy: the floors of the three-room apartment were covered in cheap linoleum, and the cement walls had never been painted or wallpapered. A clothesline ran down a dark, musty central corridor.

Chairman Wang's husband, a skinny man with a head of white hair and a stubbly gray beard, greeted me softly. Though they were about the same height, he had a smaller frame than his big-boned wife, and Chairman Wang later noted that she had always been about fifteen pounds heavier than him. He took me into the sunny room on the right, where they entertained, ate, and slept. A queen-size bed, wardrobe, and desk took up most of the space. I was unsure of where to sit. "Sit on the bed," he said, patting the comforter. He insisted that I wait there while he helped his wife prepare lunch. Their son, who was a year younger than me and was at work as a security guard, lived in the other room with his girlfriend.

When the food was ready, we improvised a dining table out of two benches pushed together and covered with a large plastic placemat. We squatted on short stools and ate tofu, stir-fried mushrooms, a fried egg, and a bowl of rice. I noticed that all the dishes had been cooked with much less oil than what the chefs used at the cooking school.

"Home-style cooking is different from the banquet-style dishes we teach at the school," Chairman Wang explained. "I don't use much MSG, maybe just a sprinkle in the mushrooms." It was pleasant eating there, with the sunlight streaming onto the bed and the plants.

After that, I visited the Wangs regularly, and they began to let me cook in their kitchen, teaching me to prepare the dishes they ate at home. The food was the perfect antidote to the fancy dishes I was learning to make. The Wangs' kitchen was barebones—more the shell of a kitchen than an actual one. Rather than working at countertops, the Wangs squatted at a low table to chop and prep ingredients. The stove, which was hooked up to a waist-high propane tank, had two burners. The tank stood near the window, which was always kept open, in case of a gas leak. Two refrigerators sat on either side of the corridor, just outside the kitchen: a fancy new silver model that the Wangs had bought before Chinese New Year and a shorter green box, their first fridge, which they had bought in 1986 and couldn't bear to throw away. They used the old fridge as a pantry, stuffing it with condiments, fruit, and packages of dry goods. Like many Chinese who had lived through tough times, the Wangs hoarded all things edible, from the little chili packets they got with airline meals to the free samples of chocolate wafers at Carrefour.

I had felt like a nuisance to Chairman Wang when I first started at cooking school, but now that I visited her at home, she seemed to take pride in teaching me. "Young people don't know how to cook anymore," she said one evening as we ate. She pointed to the room that her son and his girlfriend occupied. "They don't know how to cook. Young couples visit one set of parents to eat a few days a week, and then eat at the other parents' home on the other days. They don't even know how to cook the basics. It's sad."

STIR-FRIED MUSHROOMS

¼ pound shiitake mushrooms
¼ pound enoki mushrooms
¼ pound oyster mushrooms
     (button mushrooms may be substituted)
1 tablespoon vegetable oil
½ teaspoon minced garlic
1 tablespoon rice wine or sherry
2 teaspoons soy sauce
¼ teaspoon salt, or to taste

Cut off the stems of the shiitake mushrooms and discard them. Slice the caps into ⅛-inch pieces. Trim the brownish ends of the enoki mushrooms and separate them into smaller bunches (you don't have to separate each strand individually). Cut the oyster mushrooms into bite-sized pieces. (If using button mushrooms, discard the stems and slice the caps into slivers.)

Place a wok over high heat and add the oil. When the oil is hot, add the minced garlic. When the garlic begins to sizzle, add the shiitake and oyster mushrooms and stir-fry for 3 minutes. Add the rice wine and soy sauce, then the enoki mushrooms. Add the salt, stir, then reduce the heat to medium and cook for 5 minutes. Remove from the heat and serve immediately.

3

EVEN AS I LEARNED to cook privately with Chairman Wang, I continued to go to my regular classes at the cooking school. Though Teacher Zhang dispensed a fair bit of nonsense, as usual, I noticed that my time in the classroom was helping my written and spoken Chinese. And while I didn't get much hands-on practice, I never got bored watching the chefs perform their magic. Plus, I still enjoyed lunging for the free samples at the end of every class.

I became less of a novelty among my classmates. The stalker chef, Tie Gang, began ignoring me after he realized his advances weren't getting him anywhere. Teacher Zhang made less fun of me as my Chinese improved. And another woman enrolled at the school.

Mrs. Zhao looked like a middle-aged librarian with her large eyeglasses and short, permed hair, but she acted more like a spoiled schoolgirl. On her first afternoon in the kitchen, she sucked on a bottle of yogurt through a straw, and strutted around boldly interrupting whomever she liked. Chairman Wang and I were finishing a private lesson, and students were filtering in for the demonstration class. I smiled at her, but she didn't seem to notice.

She made it clear that she didn't know a thing about cooking and that she was here just for fun, thank you very much.

"Chairman Wang, I think I'm going to have a hard time in this class," Mrs. Zhao said.

"Not necessarily," said the chairman in her best deadpan.

Mrs. Zhao paused to watch me as I chopped my pork. "Actually, I think I can do better than her!" She resumed pestering Chairman Wang. "If I pay my tuition next time, will I be able to participate?" she purred. "Chairman Wang, what's your phone number? Can I call you? Where can I buy a knife? Can I buy one for twenty-five dollars?"

"Four or five dollars is enough," the chairman said curtly, and from that point on she ignored Mrs. Zhao. Undaunted, Mrs. Zhao began flirting with a student.

Mrs. Zhao continued to irritate me in the classes that followed. She flaunted her status as a housewife, a rare position in a country where most women couldn't afford to stay home. She drove her black sedan to school and parked it near the bicycle stands. She fussed over the amount of MSG the teachers used in their dishes. But I couldn't ignore the reality that I had more in common with her than I did with the other students. Didn't I also bombard the teacher with questions? And wasn't the class basically a hobby for me, too?

I didn't *think* I was as annoying. For the most part, I took pains to hide the differences between my classmates and me. I dressed in old jeans and a fleece pullover unless I had to be somewhere important after class. Aside from the time I had flashed my passport—and perhaps the one occasion when I'd worn a suede jacket—I tried to keep a low profile. Then I thought back to another afternoon when I'd hauled my Apple iBook, without its case, to class, having just picked it up from the repair shop, and another time when I'd brought a foreign friend who, with her fair complexion and blond hair, stuck out like a rock star. Finally, I had to admit it to myself: I was more of a showoff than Mrs. Zhao, whether I liked it or not.

After ignoring me the first day of class, Mrs. Zhao in subsequent classes tried to befriend me. Annoyed and uncomfortable with her, I declined her invitations to visit her home. Though she sat next to me on the bleachers, I didn't ask her for help with my notes. When she called to ask if I could bring her back some vitamins the next time I went to the States, I told her it would be very difficult. That was how you let someone down the Chinese way.

A month or so into cooking school, I decided that I was going to take the national cooking exam along with the rest of the students. Taking the test would give me a goal, an intellectual challenge that I hadn't had since my school days. For me, it was a matter of pride. I wanted to prove that I was just as capable as the rest of the students, even though, as Teacher Zhang had put it, Chinese wasn't my mother tongue.

I wasn't too worried about the cooking component. I had mastered a fair number of dishes during my private lessons with Chairman Wang. I could easily steam a fish with ginger broth, dry-roast string beans with minced pork, and braise tofu in a spicy Sichuanese sauce. What worried me was the written part. It supposedly tested students on how much we had learned about cooking fundamentals in the classroom; but given the way the questions on the sample exams were phrased, I suspected we were going to be tested on how well we had memorized our textbooks, down to their minute—and often baffling—details. A typical question lifted a passage directly from one of the books and asked the student to fill in an omitted word:

A cockroach can survive in −5 degrees Celsius conditions for _____ minutes.

A. 5   B. 10   C. 15   D. 30

A couple of thoughts went through my head as I tried to eliminate each choice: Weren't cockroaches able to survive a nuclear attack? Was it common for chefs to moonlight as exterminators?

Impatiently, I flipped to the back of the test book. The correct answer was D. But coming up with the right answers was the least of my problems. When I sat down with a sample test and attempted to read it in earnest, it took me three hours just to comprehend the first twenty questions, let alone answer them. Since then, I had been carrying around my textbooks and sample tests everywhere I went, but I couldn't bring myself to repeat the agonizing experience on my own. That was when I called Chairman Wang.

Several afternoons a week in the month before the test, we sat next to each other at a rickety table she had set up in the middle of the cooking school's kitchen. The table was strewn with sample tests and my textbooks. The chairman brought her small, weather-beaten Chinese dictionary, and I was armed with a very large and heavy Chinese-English dictionary. I stumbled upon mystery after mystery:

Which of the following abilities does protein not possess?
A. It prevents edemas.
B. It makes antibodies.
C. It makes bones and teeth.
D. It keeps the brain normal and happy.

It had to be D. D sounded so absurd that I figured it must be the right answer.

"Wrong. It's C," said Chairman Wang. "Go figure. I'm not any happier after I eat protein."

She vented along with me when I got yet another answer wrong.

"*Budaga!*" she would exclaim, using Beijing slang that meant "That can't be!"

A number of questions concerned parasites, death, and fecal matter. I learned that ingesting more than three grams of chemical meat tenderizer could kill a person. I learned that I should kill an eel by drowning it in boiling water. I learned that tapeworm did not need an intermediary host to infect a human.

"Is tapeworm still a problem in China?" I asked.

"Not anymore," the chairman said. "When we were growing up, yes. I guess it had something to do with growing our food in night soil. It's considered unclean now, but the vegetables sure tasted better. Cucumbers don't have cucumber taste anymore because they're grown with chemical fertilizers."

I memorized the terms for specific cuts of pork, beef, and lamb. I had to know if each cut was fatty or lean, tender or tough, gristly or not, meaty or surrounded by bone. I memorized which cut was good for making fillings for dumplings, which for sweet-and-sour pork, and which were suitable for stock. Chinese pigs were divided into sixteen cuts, including the tail, neck, and head. Each part of the butt had its own name.

For one of the questions on a sample test, I had to fill in the blank:

The stacking method is used in dishes containing
ingredients that don't have bones, are _____, or
are crisp.

A. soft   B. very soft   C. *renxing*   D. hard

I looked up *renxing* in my dictionary. It meant "pliable but
strong; tenacious."

"Ah," I said. "People are *renxing*."

Chairman Wang liked my sample sentence. "That's right!
We can go hungry for three days and still live. People are tenacious. We can live through very, very bad times."

The exam would test exactly how *renxing* I was. I needed to
correctly answer sixty out of a hundred questions to pass.
Eighty of the questions would be multiple choice, and the rest
would be true-or-false. The true-false questions would be trick-
ier than they seemed. To save time, the chairman advised, I
should mark all these questions false. False came up more
often than true, in her experience. Then I could concentrate
on the multiple-choice questions, which were more straightforward.

She must have detected a despondent look on my face.

"Don't worry! You're not taking the *gaokao*," she said, referring to the notoriously rigorous national college entrance
exam. "Just go to your test like this." She adjusted her posture,
raised her head in a dignified manner, and tucked the textbooks casually under her arm. "Don't make it obvious. Maybe
hide them a little, otherwise someone might take the books
away from you. Then keep them on the floor or in your lap
during the test."

It took me a moment to process what she was counseling.
Surely she was kidding.

She went on with the scenario, like a bank robber plotting a heist: "You'll be given ninety minutes to take the test. After the first thirty minutes, the official sent from the Ministry of Labor will go, leaving the two proctors. One of the proctors is usually an administrator from our school. When they start chatting with each other, you'll know it's okay to cheat. I'd say that eight out of ten proctors will let you copy from someone."

Seeing the look of surprise on my face, she added, "Of course, no one should cheat." But with limited resources in the dog-eat-dog world that was contemporary China, students didn't want to take chances. The exam cost $40, and if a student failed it, he would have to pay $12 to retake it.

"Some of the students have never gotten a proper education, so they have to cheat," the chairman said. "Some are from really poor villages. Some grew up in caves. Some simply don't have the brains. So we let them copy."

Beijing used to be the promised land for migrants, Chairman Wang continued. "In the early nineties, migrants could come here and make a lot of money. Not anymore. Now everything has stabilized, and it's harder to find opportunities. Not everyone can find a rice bowl. These days, everyone's a chicken head, not a phoenix tail." A rice bowl was a metaphor for a stable job, a means of livelihood. A chicken head meant something ordinary, while the tail of a phoenix was something of high value.

"Some of the students are laid-off workers," she added. "Some have worked in clothing factories, sewing sleeves all their lives. But their factories have closed and moved down south. So the government pays for their tuition, and they're here to learn a new skill to help them find a job." The laid-off workers were allowed to choose from a number of vocational schools, which taught such skills as auto repair, flower arranging, and massage.

We went back to studying. I bombed question after question.

"Do you think I have any hope of passing?" I asked.

The chairman thought for a moment. "Why don't you hire a gun hand? Go to the administration and tell them, 'I'm working very hard.' Tell them you'd like to save some effort and have someone take it for you." A gun hand was a hired professional who took the exam in place of the student. There was enough corruption in China's test-taking industry that gun hands were easy to find.

I supposed that the chairman's response was her roundabout way of saying no. No, I had no chance in hell of passing on my own.

I was in mad pursuit of the perfect circle. I rocked the pin back and forth over the dough, creating dumpling skins shaped like a heart, the continent of Australia, and one of those squiggly blobs used by psychologists. No circle had materialized yet.

I was at the Wangs', taking a break from studying by making dumplings, and the first priority was to learn how to make the skins. When I had wrapped dumplings with my family in California, we had always used frozen wrappers. Chairman Wang allowed no such abomination in her household.

In a large tin bowl, she first combined flour with water.

How much water? I asked.

"I don't know. You've made noodles before, right? Well, more water than you would put into noodle dough. Dumpling dough is softer."

The sun was setting as we worked. The kitchen didn't have any electricity, so it grew harder by the minute to see what we were doing. Finally, as it neared pitch darkness, Mr. Wang came in with a light bulb. He draped the cord around a nail on the kitchen wall and plugged it into an outlet in the corridor. The bulb hung limply from the nail, casting faint shadows in the small room.

Chairman Wang rolled out a skin and held it up to the light. "You see this? This is a good one because there aren't any dark spots." A soft light shone through the skin as if it were a lamp-shade.

The Wangs made dumplings once or twice a week. One batch was enough to last them for at least two meals. "Dump-lings fill you up. I can't get full eating rice," said Mr. Wang. "A few hours after eating rice, I'm hungry again."

"I like making dumplings because it's less labor-intensive than making a normal meal," said Chairman Wang. It didn't seem that way to me, what with kneading the dough, rolling it out, mixing the meat filling, and wrapping the dumplings.

"Wouldn't it be easier to eat dumplings in a restaurant?" I asked.

"Yes, but they wouldn't be as good," Chairman Wang said. "Homemade dumplings are the best."

I watched as she poured soy sauce into a large tin bowl full of ground pork.

"How much soy sauce do you put in?" I asked.

"It depends," she said. "If you want more flavor, add more soy sauce. If you want less flavor, add less soy sauce. Just make sure it doesn't make the meat mixture black. That would be too much."

She sprinkled in a bit of chicken bouillon, then a good help-ing of dried shrimp. Each time I asked her for another meas-urement to jot in my notebook, she answered vaguely: It depends. If I wanted more of that flavor, add more; if I wanted less of that flavor, add less. I was beginning to realize that my American attentiveness to measurements sounded strangely obsessive to Chinese.

Wang added: "It's different for everyone. It depends on your taste buds."

The only ingredient that had to be slightly more exact was the water: "It depends on how much meat you've used. You add enough so that the meat feels springy." She added half a rice bowl's worth to start, then another few trickles, mixing the meat vigorously with her chopsticks each time she added more moisture.

PORK, FENNEL, AND SHIITAKE MUSHROOM DUMPLINGS (*ZHUROU HUIXIANG XIANGGU JIAOZI*)

3	large eggs
1 to 2	teaspoons vegetable oil
¾	pound ground pork
½	cup water
⅓	cup soy sauce
½	teaspoon salt
1	tablespoon sesame oil
2	teaspoons finely minced garlic
1½	teaspoons finely minced ginger
1	tablespoon finely minced leek, white part only
1	fennel bulb, finely diced
1	cup finely shredded napa cabbage
4	shiitake mushrooms, finely diced
¼	cup dried shrimp (optional)
80	dumpling wrappers (see next recipe)

To make the filling: Beat two of the eggs. Heat a wok to medium high, add the oil, and scramble them. Cut the eggs into small pieces and set aside. In a bowl, mix the pork and water vigorously with chopsticks or a fork, about 50 strokes, making sure to mix in one direction. The pork should have the consistency of cake batter. Add the third egg (raw) and mix another 20 to 30 strokes. Add the soy sauce, salt, and sesame oil and mix again. Add the garlic, ginger, and leek. Blend well. Add the fennel, cabbage, mushrooms, and dried shrimp (optional). Blend well. The filling is now ready to fold into the dumpling wrappers.

To wrap the dumplings: Place a dumpling skin in your palm. Scoop a dollop of filling in the center of the skin. Fold the skin in half, and pinch the top of the semicircle together. Starting from one side, pleat and pinch the edges until the filling is fully sealed and the dumpling has the shape of a crescent.

To cook the dumplings: While you are wrapping, fill a large pot with water and bring it to a boil over high heat. Add the dumplings in batches of 20, and when the water returns to a boil, cook for 5 minutes. Drain and serve immediately.

DUMPLING WRAPPERS (*JIAOZI PI*)
Makes approximately 80 wrappers

4   cups all-purpose flour
2   cups water

Place the flour in a large bowl. Stir in 1 cup of the water, then work the water into the flour with your hands. Slowly add more water, about ¼ cup at a time, mixing thoroughly so the water is fully incorporated before adding more. Stop when the dough is springy and soft, not too dry but not slippery. Transfer to a clean surface and knead for 3 to 5 minutes. Cover with a damp cloth and let sit for at least 10 minutes.

Divide the dough into three equal pieces. Roll each piece into a long rope, about ¾ inch in diameter. Slice the ropes into inch-long pieces. Lightly sprinkle the pieces with flour, and roll them with one hand on the counter to form balls. Squash each ball with the center of your palm to flatten it like a silver dollar. Sprinkle flour over the dough and work surface.

Working with a rolling pin and one piece of dough at a time, start from the center of the dough and roll outward, then roll back to the center. Turn the dough a few degrees and roll again. Continue rolling, turning the dough in the same direction, until you have made a full revolution. The dumpling skin should be flat and round and slightly bigger than your palm. It probably won't be a perfect circle the first time you try, or the second, for that matter. Your technique will improve as you go along.

Stack the wrappers and cover with a damp cloth to prevent them from drying out while you finish rolling. Use them immediately.

@

I worked on my wrappers, determined to get them right. The rolling of the skins became a quest for perfection, for just the right amount of flattening and turning and rolling. I couldn't stop; I was obsessed. The thinner the dough became, the more it felt like soft leather. The chairman collected the skins I rolled and wrapped the dumplings in a matter of seconds, then chatted animatedly while I half listened and concentrated on rolling out an acceptable handful more.

In class, Chairman Wang was fairly reticent. During the demonstration classes, she spent most of her time in the backroom, hunched over a desk reading novels. She was always courteous and good-natured with students, but formal. At home, she was more relaxed. She answered the door in her lamb's wool long johns. She grinned more, revealing a set of even white teeth. She complained about government corruption. She gossiped about the teachers at the school. Gradually, she talked more about herself.

As we made the dumplings, she mentioned that she had recently turned sixty.

"I wish I had known. I would have taken you out for a meal," I said.

"I'm not the kind of person who celebrates birthdays anyway. But I did buy myself a pair of earrings. Want to see?"

She disappeared into the bedroom and returned with a pair of gold studs. Rather than putting them through her earlobes, she fiddled with them in her palm.

"Birthdays don't have much meaning to me," she said. "I don't want to live too long. To tell you the truth, everyone loathes old people."

But wasn't it a part of Chinese tradition to respect the elderly?

"Oh, people don't really believe in that. The truth is, when people get old and become a burden, no one likes it. The old people don't like themselves, and nobody likes the old people."

The chairman had a pessimistic, even fatalistic view about many things. She believed that as China grew prosperous, society was going downhill. In the 1950s, she remembered, nobody spat on the sidewalks. Before economic reforms, people used to give up their seat on the bus for an elderly person.

"As long as corruption continues, people will act less civil to each other," she said.

After the dumplings were boiled, Chairman Wang and I set up the makeshift dining table in the corridor outside the kitchen. Mr. Wang took a plateful of dumplings into the bedroom, and I heard him clicking on the television. I dipped the dumplings in a small bowl of vinegar and chili before I bit into them. They were steaming pillows of dough with a familiar taste, but bursting with a freshness that I had not previously experienced. It was like trying handmade ravioli for the first time after a lifetime of eating the frozen version. I gobbled down a dozen. Chairman Wang polished off several more than I did. She then went to the kitchen to fill her bowl with "soup" —the water the dumplings had been boiled in.

In cooking class, we had started a unit on restaurant management and accounting. As we ate, we calculated how much each dumpling cost, based on the amount the Wangs had spent on the ingredients. The ninety-two dumplings we wrapped were made from three dollars' worth of ingredients, which worked out to about three cents per dumpling. She estimated that the average person could eat twenty dumplings in one sitting, for a total cost of sixty cents per meal.

"You'd pay just a few cents more to eat them in a restaurant," Chairman Wang concluded with a laugh. She shrugged. "I still like making them at home. They taste better, and it's cleaner to eat at home than in a restaurant."

The Wangs ate all their meals at home. I asked Chairman Wang how much they typically spent on a home-cooked meal.

"It's hard to say," she said.

"How much do you spend on food every month, then?"

"I don't know. Every month is different. Want to see?"

She shuffled into the bedroom and returned with a little blue notebook. Each page was headed with a month. Underneath each month was a series of numbers written in pencil. She had painstakingly recorded all of their expenses, down to the penny. Between Chairman Wang's job at the cooking school and their combined pensions, the Wangs earned almost $400 per month—putting them solidly in China's middle class.

They spent between $170 and $350 on their household expenses each month, most of it on groceries. In fact, the amount they spent on groceries was more than most Chinese earned each month. They didn't have many other regular expenses, though. They had bought their home with cash, so they didn't have rent or a mortgage. Because she was taller and had a bigger frame than most Chinese women, Chairman Wang sewed her own clothes, and had, until recently, stitched her own shoes. When her bicycle broke, she repaired it herself. Aside from groceries, their major expenses were for monthly bus passes and utilities.

"Have you always kept track of your money like this?" I asked.

"No. It wasn't necessary when I was making five dollars a month"—before the economic reforms. "I started doing this during SARS, when I was bored and there wasn't anything to do." The school, like many institutions, had closed during the outbreak. The chairman spent her days in the supermarket, scanning the aisles. "I tried all sorts of food. Then I figured that since I was spending money, I might as well keep track of what I was spending."

Recently, the Wangs had decided to stop saving altogether and to splurge on big-ticket items: $350 for a digital camera, $300 for the new refrigerator, $800 on a trip for two to Hainan, an island off China's southern coast.

They had also recently been duped out of $80 when Mr. Wang went to the doctor complaining of a dull pain in his side. An unscrupulous doctor, eager to make money, had recommended some unnecessary treatments, including a CT scan, an x-ray, and a number of pills.

Health care was terrible, Chairman Wang said, shaking her

head and frowning. Though everyone was supposedly covered in China, health insurance kicked in only after an individual had spent more than $150 per year. Benefits were capped at $50,000 a year. China was theoretically a socialist system, but the old welfare net was falling apart so rapidly that in some ways it was becoming more capitalist than the United States.

"If you have cancer, your costs will be more than that. So if you're really sick, you might as well just die. Don't go to the hospital."

Most Chinese I knew kept their money in savings accounts, while some had begun investing in real estate or stocks. When I asked her why she didn't do either, she explained, "We have only a little money, so we might as well enjoy what we have. Our son doesn't need our money. He'll inherit the apartment. And we've suffered enough in our lives. We've been through enough hardships. When my husband's mother and my parents were still alive, the six of us lived here. Sometimes we could afford to buy only five eggs. Everyone but me got one. I couldn't bring myself to eat one. I was stronger than everyone else anyway."

So the Wangs, who had endured China's tumultuous past and managed to make it into their golden years, were throwing caution to the wind. There wasn't any point in trying to save money to pay for a potentially life-threatening illness. They took a holiday once a year and spent the rest of their money at outdoor wholesale grocery markets, giant supermarkets like Wal-Mart, and streetside food stands.

"Some people have a different way of thinking," Chairman Wang added. "They would look at me and say that I spend a lot of my money on food and that's a waste. They'd rather buy a vase because you can put it on your shelf and make your house look nicer. They like it because it will always be there. But I get a lot of enjoyment out of eating. I like to try different kinds of food. I buy whatever pleases me."

She pointed toward the darkened corridor, unadorned except for a shelf filled with a stash of goods, including a bag of instant powdered almond tea, a loaf of bread, and a package of Sichuan-style spicy crackers.

"These are the things that make me happy," she said. "What does it mean to be wealthy?. To be able to eat, drink, and move about. That is my definition of wealth."

4

CHAIRMAN WANG WAS sixty years old. That meant she had been born right before the Communists rose to power and had lived through the upheaval of Mao's time. I was dying to know what it had been like to witness and participate in his attempts to create a Communist utopia, but I could never find an appropriate time to bring it up. Few Chinese talked about the past; often it seemed as though they didn't dare think about it. I did not want to upset her, and the formality of the teacher-student relationship made it harder to ask questions. But the more dumplings we wrapped together, the more comfortable we became with each other. It was as if this traditional family activity had a power of its own that freed us from our constrained roles.

One afternoon, Wang visited my home. Though my kitchen wasn't much bigger than the Wangs' workspace, the tile walls, cupboards, and overhead lights made it more modern than the Wangs'. We stood rather than squatted, and kneaded the dough directly on the black countertop to make lamb-and-pumpkin dumplings. As I sliced the pumpkin and Chairman Wang mixed the ground lamb, I asked her what life had been like during the Cultural Revolution. She said simply, "We were working on the revolution."

"What do you mean?" I asked.

She shrugged. "There was no meaning."

I rephrased the question. "What exactly were you doing?"

"I was writing revolutionary slogans on banners," she said as she added water to the lamb and continued to mix. "I had just graduated from high school when the revolution began. I was still boarding at the school, and none of us were allowed to leave. Our job was to stay put and 'revolutionize.' If I had sneaked out to go home, I would have been branded a counterrevolutionary."

She stared out the window as she talked, her eyes somewhere far in the distance. "That period was very dark. People were horrible! Teachers told students to beat up other teachers who weren't fervent enough about the revolution. In some places, teachers were killed."

She stopped mixing the lamb, and her voice, normally gentle and even, trembled. "At my school, there was one teacher from an upper-class family. Her parents were intellectuals. She had just had a baby, and she was supposed to be eating better food than the rest of us in the cafeteria. But because she was eating a little better, the Red Guards used that as an excuse to punish her. They shaved her head into a *yinyang* head. Do you know what a *yinyang* head is?"

I shook my head.

Her voice grew into a shrill wail. "They shaved one side of her head so that it looked like a yin-yang symbol! They made her wear a sign that listed her crimes as a counterrevolutionary. Kids younger than me would take turns beating her with a belt. They denounced her in front of big groups of people."

Chairman Wang was practically shouting now, her usual dead-pan gone, and her forehead furrowed with emotion.

I stopped slicing the pumpkin, unsure of how I should react. My gut instinct was alarm—what if the neighbors overheard her? Rationally, I knew there weren't any consequences for the chairman to speak privately in my home, but it brought out the mild paranoia that always floated just under the surface of life in authoritarian China, even if it was an authoritarian state in reform. It kept you from talking about anything sensitive. It kept you in your place.

Having already crossed that psychological threshold, however, Chairman Wang was in no mood to stop. "I could have been a Red Guard if I had wanted to be one," she said. "I was from the right kind of family. But I couldn't bring myself to do what the others did. I just tried to stay quiet. If everyone moved to the right, I moved to the right. If everyone went to the left, so did I. I wanted to become a doctor, and I had tested into the right school. I was supposed to start the following year. But then the revolution happened and all the schools were shut down. I lost my chance to go to college.

"China lost two generations. The generation above me was persecuted if they were scientists or intellectuals. And then my generation—we didn't get a proper education in the first place."

Chairman Wang should have been among the people who benefited from Mao's reforms. She didn't come from a family of intellectuals, or one that had ties to the former Nationalist government. Her parents were workers, the "correct" class that the Communists were supposed to help. But instead of becoming a doctor, she had ended up in a rundown cooking school, performing lowly tasks that none of the chefs would do.

I wanted her to go on, but we were both overcome with emotion. A very Chinese instinct kicked in, and I stopped asking questions. "Well, should we start shredding the pumpkin?" I asked.

LAMB-AND-PUMPKIN DUMPLING FILLING
(*YANGROU NANGUA JIAOZI*)

⅔   pound ground lamb
½   cup water
⅓   cup soy sauce
1   large egg
½   teaspoon salt
1   tablespoon sesame oil
1   teaspoon minced garlic
1   tablespoon minced leek
½   teaspoon minced ginger
2½   cups grated fresh Dickinson pumpkin (see note)

In a bowl, combine lamb and water and mix vigorously with chopsticks or a fork, stirring in one direction for 50 strokes. Add soy sauce and mix for another 50 strokes. Beat in the egg, salt, sesame oil, garlic, leek, and ginger. Add the pumpkin and stir for another 10 strokes. Wrap and boil as described in the dumpling wrapper recipe in the previous chapter.

Note: Dickinson pumpkins, oblong and yellowish, are found in farmers' markets across the United States. Butternut squash makes a good substitute. Do not use a round, orange Halloween pumpkin, which lacks the proper flavor and texture.

We stopped talking about the past that day in my kitchen, but only temporarily. The next time we met, Chairman Wang picked up where she'd left off, and after that it was impossible to stop. I couldn't stop asking questions, and she couldn't stop recalling. It was as if a crack had appeared in a floodgate, a crack that was impossible to seal up. Her story leaked out in small bursts, bit by bit, every time we met to cook or study.

Winter turned to spring, and the temperature in the hallways of the Wangs' apartment building became almost bearable. The Dragon Boat Festival was a few days away. When I knocked on the Wangs' door, the chairman answered in a pair of light pants and a bra. The bra didn't look like any I had ever seen. It was gigantic, a hybrid between a vest and a brace, with a row of buttons down the front. The straps covered her wide shoulders. I hadn't noticed before that Chairman Wang was a well-endowed woman.

"Ni hao," she said, closing the door behind me.

We squatted in her kitchen wrapping *zongzi*, a traditional Dragon Boat dish. We took reed leaves, folded them into a cone, stuffed them with raw glutinous rice and raisins, and folded the leaves over the top, tying up the dumplings with string as if they were gifts. We made dozens of the small green parcels, each shaped halfway between a cone and a pyramid, dangling from strings. The chairman filled a wok with water, unscrewed the knob on the gas tank, and fired up the stove.

Chinese associate almost every holiday with the eating of a particular dish, just as Americans have Thanksgiving turkey and Fourth of July barbecues. The chairman told me that the Dragon Boat Festival commemorated the death of a prominent poet named Qu Yuan, who lived in the third century B.C. Despite being an honorable and honest ally to the government, Qu fell out of the emperor's favor, and he was banished from the kingdom. In despair, he drowned himself in a river. When nearby villagers got word of his death, they rushed out in boats (called dragon boats because ancient Chinese believed that dragons ruled the waters) and threw rice into the river to pre-vent the fish from eating his body. In modern times, *zongzi* came to symbolize the rice.

As the dumplings boiled, Chairman Wang asked if I wanted to see something from her past. She went into her son's room, rummaged through a cabinet stuffed with junk, and pulled out a rainbow-colored mobile. Hanging from the strings were twisted pyramids resembling *zongzi*.

"I made this in high school," she said, dangling the mobile in the air, "when there wasn't much to do."

As a young man, Chairman Wang's father had worked as a la-borer on a British ship that moved goods between Asia and Europe. "I remember him talking about places like Marseilles," said the chairman. "He spoke English fluently, but he couldn't read a word of it."

After her father married, he became a bicycle repairman. Her mother was a typical housewife. She was born in 1912, one year after the Qing Dynasty was overthrown, ending thou-sands of years of imperial rule. Women no longer had to bind their feet or serve as concubines, but many women still could not read in the early years of the republic. It was only after the Communists came to power in 1949 that women's literacy and employment became priorities. Wang said, "After Liberation"— the Chinese term for the founding of the People's Republic— "Chairman Mao instituted 'illiteracy elimination teams.' The teams went into neighborhoods and taught women like my mother how to write her name and read some basic things." It was one of Mao's lasting contributions, she added.

Chairman Wang's mother gave birth to a boy first, and a few years later to Wang. One of the childhood activities Wang remembered was called "forging iron and steel." In the late 1950s, Mao commanded the entire population of China to increase the country's steel production. After school, Wang and her classmates wandered around looking for rusty nails and scraps of steel lying in the dirt roads. "The teachers praised whoever found the most iron and steel, which was the only motivation we needed to go out and find more." Iron woks were confiscated and melted down, and people were sent to communal canteens to eat. The canteens lasted only for a week or two in Beijing, Wang said, but the madness of iron production continued. A huge pit had been dug on the school's campus to mine any iron that might have been embedded underground. No one found much iron there, but the school kids enjoyed playing in the dirt.

"We didn't have any proper toys," she said. "We had rubber bands. We made dice out of the joints of sheep. We made toys out of paper."

The chairman also recalled playing on Beijing's old city wall. "The wall marked the edge of the city," she said. "I remember seeing opium addicts loitering near the wall." Living within the gates, Chairman Wang was sheltered from the problems of the countryside, including the massive famine that struck China in the early 1960s. She showed me photographs from her adolescence in which she wore thick eyeglasses, two braids, and a padded jacket. After having heard so much about starvation in China during that era, I was surprised to see that she was plump and well fed; it turned out that most of the capital had largely been spared from the tragedy.

"I was lucky. Our family never had a problem getting food," she said, acknowledging that this wasn't the case for many people in China. The issue hadn't been quantity, but quality.

In her youth, noodles and vegetables were plentiful, but meat and eggs were considered luxuries.

Mao officially launched the Cultural Revolution with a series of demonstrations in Tian'anmen Square in 1966. Chairman Wang joined the hundreds of thousands of students who waved copies of Mao's Little Red Book at the rallies.

"I was nineteen. I was really excited, but I later realized that the feeling didn't come from within. It was an infectious feeling from the crowds," she said. "Everyone around me was so moved that I was, too."

Chairman Wang didn't consider herself a Red Guard. The Red Guards were loosely formed gangs made up of kids from the right kind of background: their parents could not be intellectuals or former landlords (private land and homes had already been taken away) or have ties to the Nationalist government, which had fled to Taiwan. These ragtag gangs, empowered by Mao's imperative to overturn society and urged on by the Communists, destroyed temples, books, musical instruments, vases, and anything else that was considered "bourgeois." They intimidated and attacked teachers and those deemed "capitalist roaders." Sometimes they beat the victims to death. The gangs occasionally split into factions and waged war against each other.

Although the chairman managed to keep her distance from the Red Guards' brutal activities, as a student she was compelled to participate in certain events with the requisite Party fervor. Western-style dancing was banned, but revolutionary dances were, for a short period, performed three times a day before meals. She demonstrated one afternoon after we had finished eating in my apartment. She pretended to hold a Little Red Book in one hand and swung her arms and legs in jolting, machine-like motions. She tried to keep a straight face but burst into laughter.

"I never thought I could look back at history and laugh," Chairman Wang said. "We certainly never thought it was funny at the time, even though it was completely absurd. We were like robots." She corrected herself. "Actually, we were in a loony bin."

In the summer of 1967, with China in chaos, Chairman Wang traveled around the country in the *Da Chuan Lian*, the Big Red Guard Link-Up. The Communist government made travel by train free for students, so they could advance the Cultural Revolution in remote towns and in the countryside. Chairman Wang and her three best friends didn't care about proselytizing, but they liked the idea of traveling for free, so they headed for the train station.

"You would go to the booth and ask for a ticket to anywhere you wanted to go. If they didn't have any tickets available for that place, you'd ask for tickets to somewhere else." They traveled to several towns west of Beijing, including Xi'an, an ancient capital where the Terra Cotta Warriors were later discovered. When a train pulled into a station, students had to clamber in and out of the windows because the carriages were so crowded. "If you were standing on the train and you lifted your foot, you'd have nowhere to put it down." After a few weeks, the fun had worn off. They couldn't get a spot on a train back to Beijing, and one of the girls had gotten sick. Finally they met an official who helped them forge a letter that got them on a train back home.

After two years of chaos, Mao quelled the Red Guard movement. But most of the schools remained closed, and millions of urban youths—including Chairman Wang—were "sent down" to live in the countryside and do manual labor with the peasants. The service was compulsory and indefinite. "The government didn't tell us how long we'd be gone for, and we didn't ask," she said. "I thought it was quite possible we would be in the countryside for the rest of our lives." As consolation, perhaps, the students had some say in where they were sent. Chairman Wang didn't want to go north, to Manchuria, because it was too cold. Tending horses and sheep in Inner Mongolia didn't appeal to her either. Yunnan, in China's far southwest, was several days' travel away, so she decided to join a trainload of youths bound for Shanxi, a poor province just west of Beijing.

I was surprised when she spoke fondly about her time toiling in the fields; she practically made it sound like an extended summer camp. "There were thirty of us in my production brigade, and the locals built us two rows of houses. Three of us lived in each room. We grew wheat, barley, cotton, millet, and sorghum. Shanxi soil wasn't very good. It was loess, and we were on a mountain plateau, so we couldn't grow rice. For two years, I didn't eat rice. I didn't really like millet—to this day, it makes me want to throw up when I think about it. We ate a lot of *wotou*"—steamed cornbread in the shape of an inverted cone. Years later, Chairman Wang still steamed up a batch of the rural staple occasionally, for old times' sake. "We ate idle meals," she said wistfully, using an idiom that meant something like "Those were the days."

"But there was also plenty to do. We worked in teams of two, spreading seeds. Once the seedlings began to grow, we had to thin them out. One person would push the dirt with a hoe, and the other person would stoop down to arrange the seedlings. Then the person with the hoe would push the dirt back over them.

"When we weren't planting, we were digging ditches, mining coal, and fetching water from the well. We built roads. We sang Cultural Revolution songs." She belted out:

Chairman Mao created a just war in the east,
Spring, summer, autumn, winter . . .

She paused. "Oh, I've forgotten the rest! The villagers were nice to us. Because we were from Beijing, they thought we knew Mao personally, so they were afraid to treat us badly."

In lieu of a salary, the students were given work points: ten points per day for men, eight points per day for women. Once a year, after all the cotton, wheat, and other grains were sold, the villagers would share the profits with the production brigade. "It worked out that each point was worth about one *fen*. I basically made enough to buy one postage stamp a day!"

After the first year, Wang actually owed her brigade $2.50 because she exceeded her food ration. "I could eat four *wotou* a day. I was stronger than most of the women, and I worked harder." The next year, she worked enough to repay the debt and make $2.50.

After two years on the farm, a work unit leader took her aside one day. "He asked me to evaluate my parents and myself. I told him that my father was a bicycle repairman. I didn't say whether I was a good or bad worker. I told him, 'If you want to know about my character, ask my comrades. I can't say for myself.' I had no idea why he was asking me those questions." A month later, the work unit leader congratulated her. She was going to be sent to a nearby copper mine to work as an engineer's apprentice. She didn't have a choice, but she was glad; she had grown weary of the manual labor. Others in her brigade later left — some were sent to a camera manufacturer, while a few went to a manure factory. A handful stayed behind and married locals.

The copper mine where Chairman Wang was sent needed engineers. But since the universities were closed, engineers had to be trained on the job. She made the equivalent of $2 a month during her apprenticeship. The following year, she was promoted to a full-time engineer's position, at a salary of $5 a month.

"I learned everything in one year. My responsibility was to look after the filtering equipment at the plant that provided electricity to the mine. Coal would come down on belts, and it would be burned and converted into electricity. I had to make sure the electricity levels were stable. Our job was to constantly check the meters, to make sure that the needles were pointing to the center."

For six hours a day, her gaze roamed around a room full of gauges. "If the needles started swinging, we would have to check on the coal and make adjustments. It was a very stressful job. If something malfunctioned, all the machinery would have stopped working in the mine. The miners wouldn't be able to see. It could have resulted in life-or-death situations."

Wang had an assistant, a young woman from a town near Beijing. Less than a year earlier, this woman had gotten pregnant while working at the power plant. "She and a young man from her hometown were often seen walking together as a pair," Wang said. "We each had about an hour of privacy in our rooms every day because of how our shifts were organized. That's how the pregnancy must have happened."

No one noticed as the assistant's belly swelled; it was the middle of winter, and everyone wore thick clothing. Then one day when she was in her dormitory room alone, she went into labor. She managed to deliver the baby by herself.

"Then she suffocated it and put it in a box beneath her bed," Chairman Wang said.

After someone discovered the corpse, the leaders of the plant debated how to punish the woman. But her father was well connected in the Party, so in the end she wasn't punished. The biggest problem was that no one wanted to work with someone who had such a tarnished reputation. The chairman took the woman on her shift, feeling sorry for her.

"She wasn't a bad person. She was young, and there was so much pressure at the time," the chairman said softly. "She later married the man, and they eventually had another baby."

Not long after Chairman Wang took the young woman on as her assistant, a large piece of coal got lodged in a funnel. The assistant used a stick to try to push it through. The stick broke in half, and the machine broke down. The chairman managed to alert the mine in time for the workers to halt the production line. After the machine was fixed, the boss came around and asked what had happened. Chairman Wang, knowing that her assistant had already gotten herself in enough trouble, took the blame. The boss seemed to suspect that it was the assistant's fault, but he told the chairman to write a self-criticism and let the incident go.

STEAMED CORNCAKES (*WOTOU*)

1   pound yellow cornmeal
½   teaspoon baking soda
¼   cup light brown sugar
2   cups water or milk

Place the cornmeal, baking soda, and sugar in a bowl. Mix in the water to make a smooth dough. Break the dough into pieces of about ½ cup each. Shape each piece into a cone with your hands. Place the cones, pointy tips up, on a steamer rack (sold in cooking supply or housewares stores alongside the woks) over a wok filled about halfway with boiling water, and steam for 20 minutes.

◎

Seven years after being sent to Shanxi, Chairman Wang found a way to return to Beijing. She discovered there was an official loophole, called *kuntui*—one could be excused from the countryside because of family difficulties.

"My parents were old. My brother wasn't in Beijing to take care of them," she said. "So they allowed me to go back."

She was transferred to the bicycle factory where her father worked. It was a rare opportunity then; of the twelve million urban youths "sent down" to the countryside, she was among the first to return to her hometown.

Not long after she arrived in Beijing, Wang began contemplating marriage. She was in her early thirties. It had been hard to think about such things during the Cultural Revolution, so far away from home and living without privacy. Now that she was in Beijing and the revolution was waning, she could get on with her life.

A coworker at the bicycle factory wanted to fix her up with a man also surnamed Wang. He was an elementary school teacher, and one of his students was related to Wang's coworker. The coworker was straightforward with Chairman Wang: the man didn't have a lot of money, but he was decent. Would she be willing to meet him?

"What was your first impression of him?" I asked one afternoon. Chairman Wang was sitting on her bed with her foot in a cast. On her way home from work a few evenings before, she had tripped on the freshly paved sidewalk in front of her apartment. Mr. Wang, a dutiful husband, had gone to the cooking school to fill in for her, and I had brought her a container of *zhajiang* noodles that I'd made at home.

Mr. Wang was *hai xing*, she said. "Not bad" was the same expression she used to describe my cooking. "It certainly wasn't love at first sight," she added.

When I had flipped through their photo album, I had seen a young, handsome Mr. Wang sitting on a boat on one of Beijing's lakes. Why was she describing him so unromantically? I asked.

"It wasn't a romantic time! We had just gotten out of the Cultural Revolution. We were ordinary people. There was no way for it to be romantic."

The parents of Mr. Wang's student arranged for the two to meet in their home, with the parents, the student, Mr. Wang, and Chairman Wang all sitting awkwardly in one room. "We couldn't say much to each other. He told me he didn't have many filial responsibilities. He said that he lived with his mother, that his father had already died. I told him that I lived with both of my parents."

The student's parents joked that since they were both surnamed Wang, they belonged together. "'Five hundred years ago, you were a family. So it was meant to be.'"

The two arranged to meet the following week, in a park at seven-thirty in the evening. Chairman Wang arrived on time, but Mr. Wang was nowhere in sight. She waited. He didn't show up until three minutes before eight. She was furious, and turned around to walk home. He ran after her, apologizing. He explained that a child at his school had been injured, and it took a long time for the parents to be found. He had to wait until they arrived before he left. Could she forgive him? He offered her his padded cotton jacket to wear in the cold, and she let him walk her home. They agreed to meet again the next week.

"And guess what happened? He was late again!" she huffed. "He said, 'I'm only ten minutes late this time. I'm a slow person. It's my biggest weakness.'"

Fortunately, he didn't have many other weaknesses. He had a mild temperament. He didn't drink or smoke. Aside from his tardiness, the only thing she didn't like about him was his age: Mr. Wang was six years older than she, and one year older than her older brother. In a society where birth order and age were important in defining relationships, it posed a dilemma. It was unclear who would defer to whom.

"What would he call my brother?" she said. "'Older brother' or 'younger brother'?"

Mr. Wang suggested marriage three months after they met, in December 1978. "How about it?" he asked. "If we get the marriage certificate now, we can register for furniture before the year's end. If we wait, then it might take another year to get our furniture."

At their age, they didn't have to ask for their parents' approval, knowing that their parents would simply be glad that their children were marrying at last. Mr. Wang bought ten *jin* of thinly sliced lamb, and the families celebrated the engagement over Mongolian hot pot. A few days later, they were married on the grounds of Mr. Wang's school, in front of his fellow teachers. After the ceremony, Mr. Wang returned to his classroom to teach, and Chairman Wang went back to the bicycle factory to work.

BEIJING-STYLE NOODLES (*ZHAJIANG MIAN*)

½   pound pork belly, cut into ½-inch cubes
2   teaspoons minced leek or scallion
1   teaspoon minced ginger
1   teaspoon minced garlic
1½   cups yellow soybean paste (preferably Liubiju brand)
½   cup water
¼   cup soy sauce
     Fresh or dried Chinese wheat noodles
½   cucumber, shredded
¼   cup minced parsley

In a wok, stir-fry the pork belly over high heat. (The fatty pork should not need any added oil.) When the meat begins to render its fat, add the leek, ginger, and garlic. Continue cooking until the meat is browned, then remove the wok from the heat and transfer the contents to a bowl.

In a separate bowl, combine the soybean paste with the water, stirring until the paste dissolves.

Place a clean wok over medium heat and add the soybean paste mixture. Cook until the mixture thickens and turns sticky, scraping the bottom of the wok with a spatula to keep it from scorching. Stir in the pork. Reduce the heat to low and continue to stir until the mixture is glistening and caramelized, about 10 minutes. Stir in the soy sauce and cook for another 1 or 2 minutes. The sauce can be prepared in advance up to this point. It can be served either hot or cold and can be stored in the refrigerator for up to two weeks.

Boil the noodles until tender. Drain, top with the sauce, and garnish with the cucumber and parsley.

5

As we wrapped up my last study session before the cooking exam, Chairman Wang smiled confidently as she gazed at me through her thick glasses.

"You'll do fine," she said.

After many hours of drilling me on everything from preservation techniques to steaming methods, it was nice to know that she had gained some faith in me. I was touched, until she added, "If you have a problem during the test, ask your classmates for help." Or I could ask President Zhang of the cooking school, who would be at the testing center to make sure everything went "smoothly." Chairman Wang assured me that she had told everyone about my "situation."

I sighed. Maybe I *should* just cheat, I thought. Everyone expected me to. Trying to be honest was beginning to feel exhausting. But then, what would be the point of taking the exam? Unlike the other students, I didn't need the certification; my livelihood didn't depend on it. I needed—wanted—the respect of Chairman Wang and everyone else. And I wouldn't get it unless I passed the exam on my own.

The test was supposed to take place on two separate days. We would take the written portion first, and a week later we would be tested on our cooking skills. Initially the written test was scheduled for a Wednesday, but then we heard it had been pushed back to Sunday. Then, about a week before the test, we learned that it had been moved up to the coming Friday.

On Thursday, I called the cooking school to confirm that I had the correct date. Yes, said President Zhang, a stern woman I'd met only in passing. She sounded aloof and annoyed.

"And can you confirm the exact location of the test?" I asked.

"Isn't it written on the back of your registration form?"

"Yes, but I just wanted to be sure I understand specifically where that is. Is it on the east side of the street or the west?"

"I don't know. Why don't you go look for it today? Oh, and by the way, the cooking part is going to be held the day after tomorrow."

I protested. Two days wasn't enough time to prepare. And just when had she planned on telling us this?

"I'm telling you now," she said.

"But only because I called you! If I hadn't called, when would I have found out?"

"I am going to call all the students today," she said testily.

Any other student would have simply accepted the news, thanked the president, and said goodbye. I knew I was making her lose face, but I continued. "I can't take it so soon. How are we supposed to prepare in less than forty-eight hours?"

President Zhang relented, though her tone was icy. "All right," she said. "I'll make a special exception for you. You can take it next week."

Though I had gotten what I wanted, I didn't feel pleased to have once again proved that I needed special treatment.

. . .

I didn't hire a gun hand, and I didn't sneak my textbooks in with me. I didn't expect any help from my classmates or President Zhang. But when I arrived at the testing center to take the written section of the National Intermediate-Level Cooking Master Exam, I knew that it was going to be a very peculiar experience.

President Zhang stood outside the building entrance. The sun shone on the administrator's pale face and neck, both wrinkly and rough as a lizard's. She ushered me inside without a greeting.

Waiting at the door was the proctor, a mousy woman dressed casually in jeans who appeared to be about my age.

"Here's our special case," the president said. "Can you make sure to help her?"

I didn't really need any help, I told the proctor, but I asked for permission to use my Chinese-English dictionary.

"That's fine," she said as we entered the drafty, high-ceilinged auditorium. "But there will be no cheating," she said in a forced voice as she surveyed the room. "Not even foreigners will be allowed to cheat!" I noticed that an official-looking man with a stack of papers stood nearby—someone from the Ministry of Labor, I guessed, just as Chairman Wang had predicted.

Clusters of students from various cooking and other trade schools waited at their desks for the test to begin. My classmates all sat together, several in one row and several in the row behind them. I sat nearby, making sure to leave a gap between the group and me.

"I got here at seven-thirty," one of the students said.

"I got here at eight," Mrs. Zhao said.

I had arrived at eight-thirty, as I'd been told. The test was supposed to begin at nine. As a constantly tardy person, I was always amazed at how people in China showed up early for everything. But given how many times the administrators had changed the date of the test, I supposed it was only natural that students had to be extra prepared.

As I crammed a few last answers in my head while my classmates chatted, the president came by to give us a pep talk. "Remember to help each other out," she chirped. "Oh, and by the way, the cooking portion of the test has been moved up to tomorrow." She hadn't bothered to call the students the day before, as she said she would.

My classmates nodded, without a trace of the surprise or irritation that I had felt.

"I guess I'll drive tomorrow," said Mrs. Zhao. "Does anyone want a ride?"

At nine o'clock sharp, the proctor went to the front of the room. We had ninety minutes for the test, she announced. The earliest we could turn it in was nine-thirty. With an eye to the Labor Ministry official standing by, she added, "And no cheating." She handed out the test, printed on a white sheet as thin and wide as a tabloid newspaper.

I easily answered the first few questions. Then I felt someone hovering behind me.

"How are you doing?" President Zhang whispered, practically breathing down my neck. The ministry official was nowhere to be seen.

"Fine," I snapped. She lingered behind me for a minute or two, looking over my answers. I could barely concentrate.

A few minutes later, the proctor came by. "Do you need help?" she asked. "You can ask anyone here if you need help."

On my first read-through, I was pretty sure the forty-eight

questions I had answered were correct. I marked all the true-false questions "false," as Chairman Wang had instructed. Around me, the situation unfolded almost exactly as the chairman had predicted. The proctor and the president chatted away at the back of the room. Students threw their textbooks open on empty chairs. They yapped with each other, not even taking the trouble to whisper.

"What's the answer to number ninety-six?"

"False."

"Ninety-nine?"

"False."

"One hundred?"

"False."

"Great," said Mrs. Zhao. "I'll take you guys out to lunch. Shall we go?"

My classmates stood up and left en masse, with fifteen minutes of test time still to go.

President Zhang came around again to see how I was doing.

"I have a few more minutes, right?" I said.

"Yes, take your time," she said.

Group by group, the remaining test takers left. Finally the president herself left. Now I was alone with the proctor, in silence at last.

I looked up a few final words in my dictionary, which allowed me to complete the remaining questions, and reviewed my test sheet to make sure I hadn't left anything blank before I turned it in. I had gone five minutes over the scheduled time, but the proctor didn't seem to mind. My score would be ready in a few weeks, she said.

"Will I get to see a copy of the test?" I asked.

"No. Once it goes to the officials, you won't see it again."

"But I'm curious to find out which questions I got right and wrong."

"Sorry, we can't give out the test," she said. "It's one of our rules."

I decided to suck it up and take part two of the test along with the rest of my classmates the next day. I wanted to get it over with, and I didn't want to give President Zhang any more reason to treat me like a special case.

Chairman Wang had gone over the test instructions with us several times. We were supposed to bring a knife and the raw ingredients we needed to make four dishes. We were all required to prepare the same appetizer, a cold plate of thinly sliced meats and vegetables. To save time, we were to blanch and season the vegetables at home. The test required us to make a shredded meat dish. For the remaining two dishes, we could choose from the recipes we had seen demonstrated in class. A chef from the venerable Qianmen Hotel would watch us cook and score us on our skills. We were advised to choose standard dishes and to duplicate the preparation from our class as closely as possible. We would not be rewarded for creativity or for coming up with our own interpretations of the dish.

I managed to pull myself out of bed before seven o'clock in the morning and prepped my vegetables, bleary-eyed. I trimmed my fingernails and gathered up the groceries I had bought the night before and headed out, stopping by the market to pick up a last few items. But when I arrived at the school, I realized I had forgotten to season the vegetables for the cold dish.

"May I do the seasoning here?" I asked Chairman Wang.

"There's no time," she said. Anyway, she explained, the

judge wasn't going to be sampling the cold platters. Given our lack of experience, he wasn't taking any chances on a dish that wasn't cooked at a high temperature. The cold dish was meant to show off a chef's cutting skills. It wasn't tasty, with its deli-bought meats and bland vegetables. I myself never ordered that dish in restaurants because it reminded me of tedious banquets nobody wanted to attend. The only challenge was to slice everything as thinly as possible and arrange the slices attractively around a bouquet of cauliflower seasoned with curry powder.

But it turned out my classmates had taken the mandate to do the cold-dish prep work in advance as a license to prepare everything else at home too. Plastic bags and Tupperware containers full of prechopped vegetables and shredded meat tumbled out of knapsacks. Moreover, as they set up in the school kitchen, I noticed that they seemed to be drawing ingredients from the same containers. Chairman Wang noticed too. "Isn't that against the rules?" I grumbled to her as I cut vegetables at a station across the room.

Wang looked over at them and sighed. Loudly she said, "You guys are all cheaters. You are trying to trick me, and don't think I don't know."

She sounded more resigned than threatening, however, and aside from a couple of sheepish grins, the men ignored her and continued their charade. Later, she found out the extent to which my classmates had conspired: they had pooled their money and hired one of the students to buy and prepare all the ingredients, at a cost of $6 a head.

I noticed a fish tail sticking out of a plastic bag that one of my classmates had placed on my station. I peeked inside and discovered a whole fish with a crusty, deep-fried exterior.

"I did it at home last night," the student explained proudly, "to save time this morning." He submerged the fish in a wok full of hot oil to reheat it, then drained it and stuck a carrot in its mouth as a garnish.

Chairman Wang hollered at the students, "I just want to establish something. Will you be doing your own cooking, or is just one of you going to do the cooking for all the rest?" Then she shuffled back to my station with a plastic bag in her hand. "You haven't shredded your pork yet, have you? Well, here you go. It's part of the leftovers from the group."

"It's okay," I said. "I'll cut my own."

Chairman Wang watched as I sliced my pork, pressing the tenderloin firmly against the board and sliding my knife horizontally through the meat, keeping the blade as close as I could to the board. She finally understood.

"*Ziger kao,*" she said, nodding with approval at my audacity. "You're testing yourself. That's good. Whether you pass or fail, you'll be doing it on your own."

The judge arrived. He spent most of his time in the backroom, waiting for samples of our cooking to be delivered to him. Mrs. Zhao strolled in with nothing in her hands except her car keys. Her three-year-old daughter, dressed in a pink jumpsuit, waddled behind her, coming within a foot of the stove's giant flame and lunging at one of the carefully arranged cold platters.

"No! Don't touch!" Mrs. Zhao exclaimed. She laughed as the girl teetered away. "Actually, my daughter doesn't eat just anything. She's very picky."

"If only we could have been picky when we were young!" hooted President Zhang, who had come over to watch us cook and to fill up a tray with samples of our cooking for lunch.

Halfway through the test, Mrs. Zhao washed a few dishes, rounded up her daughter, and called out goodbye.

"Where are you going?" I asked.

"I have some business today. They'll finish my cooking for me," she said breezily, tossing a glance at two young men in front of the stove. From the backroom, the judge watched her go, apparently unfazed.

I let the students finish their communal cooking before I took my turn at the stove. I lined up the bowls that contained the ingredients for my dishes: fish-fragrant pork shreds, Sichuan-style wok-fried green beans, and sweet-and-sour pork, a dish that in China was more tangy and crisp than sweet and soggy. I had made each of the dishes several times on my own, and I felt confident that I could do them well, but I was nervous for some reason. The carnival atmosphere of the kitchen had died down, and the judge was waiting for my dishes.

"Did you remember the sugar? Wait for the oil to get hotter. Did you put in the MSG?" Chairman Wang commanded as she watched me cook, occasionally lending a hand. I decided it was too much trouble to resist her help. I shook the wok and added the MSG, though I hadn't planned on using any, and I let her scramble for the vinegar and the soy sauce. Despite her assistance, my fish-fragrant pork shreds came out with too little sauce in it.

"You'll get points taken off if the red oil doesn't seep out of it in a few minutes," said the chairman. Only in China would you lose points for a dish not being oily enough.

I tasted my string beans for doneness and added another dash of salt before Chairman Wang sent them to the judge. When I turned to the sweet-and-sour pork, the judge emerged from the backroom and watched me work, standing a few yards away with his arms folded across his chest. I charred the garlic and leek, and added the green pepper and pineapple chunks too late. As the dish came off the wok, he smiled and waved and left the room.

"He didn't stay to try the pork," I protested, as I packed up my dishes to take home with me. I had enough food to last me several meals.

"He figures you're better at cooking than the rest of them," said Chairman Wang. "He has faith in you."

The judge had leaked her some of the results: I had gotten 100 on my cold dish platter—he had found my cutting skills to be better than everyone else's—and Wang was certain I had passed the cooking portion. As for the written part, neither of us would know until the results came back from the Labor Ministry.

SICHUAN-STYLE GREEN BEANS
(*GANSHOU BIANDOU*)

 1   pound green beans
 1   quart plus 1 tablespoon vegetable oil
 1/4 pound ground pork
 1   tablespoon minced leek
 1   teaspoon minced ginger
 1/4 cup Sichuan preserved vegetable
     (*chuandong cai* or *zha cai*), rinsed and minced
 2   teaspoons rice wine or sherry
 2   teaspoons soy sauce
 1/4 teaspoon salt
 1   teaspoon sugar
 1   tablespoon water

Snap the stems off the beans and strip away the stringy fiber that runs down their sides.

Pour the quart of oil into a wok; it should fill the wok more than halfway. Place over high heat for about 5 minutes, or until a bean dropped in the oil immediately begins to sizzle. Deep-fry the green beans until little white bubbles form on their skins, about 3 minutes. Remove beans from the oil and drain.

Place 1 tablespoon of the oil used for deep-frying (strain it first if necessary) in a clean wok over medium-high heat. Add the pork and stir-fry for 1 minute. One at a time, leaving a minute between each addition, stir in the leek and ginger, the preserved vegetable, rice wine, soy sauce, salt and sugar, and finally the beans. Add the water and stir for 1 minute. Remove from the heat and serve promptly.

REAL SWEET-AND-SOUR PORK (*GULAO ROU*)

- 1½ cups plus 1 teaspoon cornstarch
- ¾ cup water
- 1 pound pork tenderloin, cut into 1-inch cubes
- 1 quart vegetable oil
- ¼ cup ketchup, preferably Lee Kum Kee brand
- ¼ cup rice vinegar
- 1 tablespoon sugar
- ¼ teaspoon salt
- 1 tablespoon minced leek
- 1 tablespoon minced garlic
- ½ cup cubed fresh pineapple
- 1 green pepper, cut into 1-inch squares

In a small bowl, combine 1 cup cornstarch and ½ cup water to make a smooth paste. Place ½ cup cornstarch in a separate bowl. Dip the cubes of pork into the wet mixture, then roll them in the dry cornstarch.

Place a wok over high heat and add the oil. Test the oil by dropping a pork cube in the oil; it should sizzle immediately. Add all the pork cubes and deep-fry until golden. Remove them with a slotted spoon and drain them on paper towels. Reduce the heat under the wok while the pork cools for 3 to 4 minutes. Then return the heat to high, add the pork cubes, and fry again for 1 minute. Remove them from the oil and drain again on paper towels.

In a small bowl, mix together the ketchup, vinegar, sugar, and salt. In another bowl, dissolve the teaspoon of cornstarch in ¼ cup water. Set both bowls within easy reach of the stove.

Place 2 tablespoons of the oil used for deep-frying (strain it first if necessary) in a clean wok over medium-high heat. Add the leek and garlic and stir-fry until fragrant, then add the ketchup mixture and simmer for 1 minute. Add the cornstarch solution and cook for another minute, stirring until smooth and thick. Add the pork, pineapple, and green pepper and stir to coat thoroughly with the sauce. Remove from the heat and serve promptly.

A month later, my test scores were delivered via text message. "Come to the Hualian Cooking School on Friday and pick up your cooking certificate," read the message on my cell phone, without a word of congratulations. I rejoiced anyhow: I had passed without cheating!

The license came in a blue booklet whose cover was imprinted with PEOPLE'S REPUBLIC OF CHINA and OCCUPATION CREDENTIAL CERTIFICATE in gold-colored ink. My passport photo was pasted inside, and another page specified that I had passed the test to become a certified intermediate chef. I had gotten 94 on the cooking portion and 72 on the written exam —12 points above the passing grade. The page was stamped with a red-ink seal from the government, and the official assessment read UP TO STANDARD.

Chairman Wang sounded proud when I called to tell her my scores. "You ate up a lot of energy for that test. The president was also very impressed with you. You were at an obvious disadvantage compared to the other students, yet you did better than some of them!"

When I shared the news with my foreign friends, they asked me what I was going to do with the cooking certificate. "So, are you going to work in a kitchen?" one of them joked.

I was vaguely aware of the realities of a professional Chinese kitchen. I knew chefs often worked fourteen-hour shifts, in squalid conditions with little compensation. I knew that women were discouraged. Once, when I had raised the idea with Chairman Wang, she had cautioned me that a certificate from the Hualian Cooking School would not adequately prepare me for a real cooking job.

"Well, then, what are we students here for? Aren't we supposed to be preparing for jobs?" I had asked.

"Most students who want to be gourmet chefs have been preparing all their lives," she said. "They go to vocational middle and high schools specializing in cooking. They're the official army. We're the equivalent of a guerrilla army. For the first few months, you'll be a dishwasher. Then maybe you'll be promoted to scaling fish and washing vegetables. And maybe, after a few years, you'll move up the ranks, but you'll never become a wok chef!"

I knew it wasn't going to be easy, but I was still intrigued by the idea of learning how to cook professionally. I wanted to infiltrate a Chinese kitchen, and I hoped that my cooking certificate would be my passport.

Not long after I found out I had passed the cooking test, I had a dumpling party at my apartment. I invited Chairman Wang, along with a dozen of my foreign friends, and she agreed to come early to help. She arrived just after seven in the evening, dressed impeccably in a blue flower-print *qipao,* an ankle-length, high-collared traditional dress. Like everything else she wore, she had sewn it herself. It was a far cry from the blue lab coat she wore in the school kitchen, and I complimented her on it.

"You look quite nice yourself," she said. She noted that she had never seen *me* looking very feminine either. We exchanged smiles. Then she asked what time I had told the other guests to arrive.

"Eight o'clock?" she yelped. "But you haven't even kneaded the dough! The filling hasn't been prepared!"

I knew that Chinese guests invited to a party that began at eight would arrive by seven-forty-five. Being late would mean you were holding up the party. "Relax," I said. "Foreigners don't come to parties on time."

But Chairman Wang would not be placated. "You really have some kind of nerve! If I were you, I would have started preparing at noon. I would never have the guts to invite ten people over for dumplings in the first place. Do you know how many dumplings you're going to have to make?"

Foreigners wouldn't eat as many dumplings as Beijingers, I assured her. And foreigners, even if they did come on time, wouldn't mind eating late.

At eight o'clock, she asked anxiously, "Where are your guests?" The first guest arrived at eight-fifteen. By nine o'clock, when a handful more had arrived, she relaxed, and we had gotten a good start on the dumplings. We made a large vat of pork-and-fennel filling, and dozens of fresh dumpling skins were stacked on the table, waiting to be wrapped and dropped into the wok of boiling water on the stove.

As my guests gathered around the table to wrap dumplings, I marveled at the power of the dish. It had been a comforting childhood ritual. It had brought Chairman Wang and me together as friends. It made a party festive. My motley crew of foreign friends formed a mini–assembly line, and soon enough they had rolled out, wrapped, boiled, and eaten several hundred dumplings. We broke into an impromptu sing-along when Boy George's "Karma Chameleon" played on my iPod, and on Rob Base's "It Takes Two," someone busted out some dance moves we hadn't seen since we were young.

Chairman Wang didn't dance or sing, but she seemed to enjoy mingling with my friends. While I was in the kitchen rolling out more dumpling skins, I heard her chatting with a friend.

"Oh, no. I've had many jobs," she was saying. "During the Cultural Revolution, I was sent to Shanxi . . ."

Just then Kanye West's "Gold Digger" came on, and I lost the rest of the long narrative that I had grown so familiar with.

⑥ Side Dish 1: MSG, the Essence of Taste

At the cooking school, one of my teachers lectured about monosodium glutamate the way she talked about salt, vinegar, and sugar. MSG was listed in the seasonings chapter of my textbook, right between soy sauce and oyster sauce. One afternoon, the teacher read the entry aloud: "Glutamic acid sodium taste essence, commonly called MSG, is a normal, daily seasoning. It comes without scent in crystal or powder form."

"But isn't MSG bad for your health?" I interrupted.

The teacher brushed off the suggestion and continued: MSG could be used in salty dishes, but not with sweet or sour ingredients. We should add it just before a dish came off the wok, at a temperature just below 200 degrees. Cooking it over a higher flame would result in a slightly rancid flavor.

In the demonstration kitchen, my teachers made abundant use of an open tin of the substance, looking as pure as driven snow. If their habits were anything to go by, I had unwittingly consumed buckets of the taste enhancer in my years of eating in Chinese restaurants—without the supposed MSG-induced headaches or heart palpitations. I had given little thought to the substance since moving to China, despite its omnipresence. I didn't use it in my own cooking, but not because I feared an allergic reaction. It just seemed unnecessary and gave dishes a salty, slightly chemical flavor. Using my friends as guinea pigs, I conducted blind tastings. Admittedly my sample size was small, but it seemed that Chinese were more attuned to the flavor of MSG, while most non-Chinese couldn't distinguish it from plain salt.

But after going through my cooking classes, I wanted to get to the bottom of the controversy. Why had I grown up with the idea that MSG was bad for my health when more than a billion people in China had no qualms about it?

To learn more about the substance, I took an overnight train to the Henan Lotus Flower Gourmet Powder factory, one of the country's biggest MSG makers. I had boarded the train in Beijing in the late evening and woke up the next morning to see a landscape of farmland and industrial buildings whizzing by the window. By midday, the train reached a small town called Xiangcheng, in the central province of Henan, a place that was considered the cradle of Chinese civilization but in modern times had lagged behind while much of the country boomed. The province wasn't known for its food—which, I supposed, made it a perfect place for a factory of taste enhancement.

After I had changed into a white robe and white hat with neck flaps and snapped plastic coverings over my shoes, a company representative escorted me onto the factory floor. White heaps of MSG covered the counters. Groups of four or five employees worked in a chain, scooping the crystals into plastic bags, weighing them, and packing the bags in boxes that would be shipped to the Middle East, Africa, and the United States. Lotus Flower sent 1,700 tons of MSG to America every year. "Some Americans think that MSG isn't very good," the company's spokesperson had told me the previous evening as we ate a flavorful meal. "But they sure use a lot of it."

Monosodium glutamate is most commonly thought of as an additive in Chinese food, but major American food companies like Campbell and Frito-Lay add MSG to their soups and chips to give them a tastier, more robust flavor. It is also found in its naturally occurring form in all kinds of foods, including Parmesan cheese, tomatoes, and cured ham.

I ran my hands through one of the white mounds. It felt like coarse, warm sand. I sniffed the granules: nothing. Later, in the privacy of my home, I sampled the raw crystals. They tasted rancid and salty. But when the substance binds with ingredients in the wok at the right temperature, it produces a flavor known as *xian* in Chinese.

I wasn't allowed in the main factory where the MSG was produced, but the Lotus Flower representative explained the process to me. The main ingredient in the company's MSG was wheat. MSG can be made from any edible starch, including corn, beets, and sweet potatoes. The starch is heated until it melts into sugar. Chemists add ammonia to the sugar and allow the mixture to ferment for several hours. The fermented broth is sterilized and then centrifuged to remove impurities. After a few other chemical processes, during which sodium hydroxide is added, the solution is reduced and concentrated under a vacuum at 140 degrees before cooling and crystallizing. The crystals are put through another centrifuge. Once dried, they are ready to work their magic in the wok.

The savory, addictive quality found in a bag of barbecue-flavored potato chips is considered a basic taste in Chinese cuisine, like sweet and salty. In the demonstration classes at cooking school, Chef Gao had scribbled comments on the blackboard next to the recipes that sounded like wine-tasting notes. He sometimes noted that a dish should have "a deep, dark color with a sweet, sour flavor" or be "clear and light, with the flavor of *xian*." *Xian* literally means "fresh," but in this context, it meant something untranslatable that could only be approximated with "delicious" or "savory." In recent years, chefs in the West have adopted the Japanese word *umami* to describe the taste. For centuries, the Chinese have used ingredients high in natural MSG to boost *xian* in their dishes. Those ingredients include soy sauce, dried fish, and seaweed.

Scientists did not know what it was in soy sauce and seaweed that accounted for *xian* flavor until, in 1907, a Japanese chemist named Kikunae Ikeda extracted glutamic acid from seaweed soaked in hot water. He determined that glutamic acid was the taste-enhancing component of the seaweed; the compound was *xian* in its pure form. The glutamic acid was combined with sodium to make it more palatable to consumers, and the following year, a Japanese company called Ajinomoto began producing the substance commercially. The company marketed MSG to Japanese housewives as a shortcut to flavoring evening meals. After obtaining American and European patents, Ajinomoto took its product abroad. In China—a country deeply responsive to *xian*—the flavor enhancer, under the name *weizisu,* was a hit.

But it irked many Chinese that a Japanese product sold

so well. By the early twentieth century, foreign powers, including the United States, Britain, France, and Japan, had set up ports and carved out settlements around the country, and anti-foreign sentiment ran high among the Chinese. The Chinese also accused the Japanese of "dumping" MSG on the Chinese market and organized unsuccessful boycotts of *weizisu*.

In Shanghai—where the presence of foreign powers was most greatly felt—a local chemist named Wu Yunchu began tinkering with *weizisu* and reverse-engineered it. He persuaded a Shanghainese soy sauce vendor to invest the equivalent of $5,000 in a factory to make his homegrown variety of MSG. They named the factory Tian Chu, Heaven's Kitchen, and changed the product's name to *weijing*, "the essence of taste." Vendors of pickled vegetables added it to their wares and offered free tastes as they pushed their carts through Shanghai neighborhoods, urging the locals to "buy Chinese." By 1929, Heaven's Kitchen was making more than 140,000 pounds of MSG each year, and less than a decade later production climbed to more than half a million pounds per year. Wu and other Chinese MSG manufacturers eventually forced Ajinomoto out of most of the Chinese market.

Ajinomoto shifted its attention to America, and from the mid-1930s until 1941, the company shipped more MSG to the United States than any other country. Campbell Soup and the American military were two of Ajinomoto's best clients. After World War II and the defeat of the Japanese, an American company began selling its own household version of MSG, called Accent. In China, Wu became a national hero for his economic defeat of the Japanese. Heaven's Kitchen's influence also expanded outside China, as the factory supplied a growing amount of MSG to Southeast Asia. Wu's MSG even gained acceptance in the United States —at an international exhibition, his product won awards, and he planned to enter the American market.

Wu's plans were dashed when the Communists took over in 1949. Wu was in the United States at the time. On his return home, he was warmly greeted by Premier Zhou Enlai himself, who proudly called him the "MSG king." Despite the praise, the Communist Party took control of his factories before Wu died, in 1953.

In the years after Heaven's Kitchen was nationalized, it reduced its MSG production and focused on making other chemicals. In 1965, China made only 4 percent of the world's MSG. Production increased steadily in the 1980s and 1990s, after market reforms were introduced and new factories like Lotus Flower sprang up. By the early 2000s, China made more than 70 percent of the world's MSG.

MSG first became controversial in America during the late 1960s, after the *New England Journal of Medicine* published a letter to the editor written by a Chinese doctor. "For several years since I have been in this country, I have experienced a strange syndrome whenever I have eaten out in a Chinese restaurant, especially one that served northern Chinese food," wrote Robert Ho Man Kwok in the April 1968 issue. The letter was headed "Chinese Restaurant Syndrome," and soon people around the country began complaining of the same symptoms Kwok had described: a burning sensation at the back of the neck, numbness, and palpitations. Scientists fed huge quantities of the substance to laboratory mice. A few years later, a report linking MSG consumption to brain damage appeared, and based on that report, officials in the federal government recommended that MSG be banned from baby food.

Though the measure is still enforced, MSG was never subject to further regulations. A 1987 study by the United Nations found that MSG was harmless for the general population and should be put in the same category as salt and vinegar. The U.S. Food and Drug Administration conducted a study in 1995 that came to a similar conclusion, though it stated that a small percentage of people, such as asthma sufferers, might experience short-term side effects.

Given how many additives appear in processed foods in America, I found it odd that there was such a backlash against MSG. But then, attitudes toward it seemed to be tied to larger events. Some historians have hypothesized that the reason MSG faced its backlash in the United States from the mid-1960s to the 1980s had to do with a widespread movement to reduce chemicals in the food chain, set off in part by Rachel Carson's seminal environmental book *Silent Spring*. It probably didn't help that MSG was perceived as something "foreign." Meanwhile, China was then recovering from one of the worst famines in history, and the Cultural Revolution was just getting under way. MSG didn't face the same backlash in China then because few could afford the expensive seasoning.

Nowadays, the cheap and plentiful flavor enhancer has begun to raise suspicion among Chinese. As they grow more affluent and food safety becomes an increasing concern, some Chinese are becoming more hesitant to use additives. Some Chinese friends of mine worry that MSG is unsafe and have switched to using powdered chicken bouillon (which, in fact, also contains MSG). On the other side of the world,

Western chefs are beginning to understand the concept of *xian,* so MSG consumption is on the rise.

A few blocks away from the Henan Lotus Flower Gourmet Powder Company, a red banner congratulated the factory for churning out 300,000 tons of MSG a year — almost a quarter of China's total MSG production. "We're almost there, though not quite," said Haihua, the company representative, as we toured the town in a nearly empty bus that could have seated twenty more passengers. She had told me proudly that this was the vehicle they used for all their VIPs. I found it more wasteful than impressive.

The ride offered a view of all the good that Lotus Flower had done for the town. The company employed ten thousand people, making it the rural area's biggest employer, and signs of the wealth it had brought were everywhere. In the center of town, residents did yoga in a new studio and went to the new cinema to see Hollywood blockbusters. Senior citizens strolled across a newly paved square, and when we turned down another street, I saw a cluster of old houses being torn down. A billboard put up by the developer advertised the new shopping mall that would be going up in their place. The same kind of development was happening everywhere in China.

After we made our way back to the factory, I interviewed Mr. Liu, a company official, in a large conference room that displayed packages of Lotus Flower MSG with English, Chinese, and Arabic labels. He'd heard about the negative publicity that MSG had generated in the States and had prepared a response. "In areas where *weijing* has been found to have ill effects, I don't think it's the *weijing* that's the problem. It's the people," Mr. Liu explained. He gave me an example:

diabetics can't eat sugar. Was it the sugar's fault? Of course not. "I personally cannot eat chili peppers. They're too spicy. But are they bad for everyone?"

Just as I was beginning to come around to the belief that MSG wasn't so bad for one's health, I got word that MSG production was bad for the environment's health—which in a roundabout way might indeed be bad for human health. A few days before I visited the factory, I learned from a reporter friend at the *Wall Street Journal* that he and other foreign journalists had been denied permission to visit Lotus Flower after the *New York Times* had run an article exposing how the plant was polluting the nearby river. Environmentalists suspected that the water pollution was causing cancer. In its effort to purify MSG into edible crystals, Lotus Flower had dumped 124,000 tons of untreated water every day through secret channels connected to the city sewage system, nearly quadrupling pollution levels, according to a government report.

"Whatever you do, don't drink the water!" my friend urged, and then asked, "How did you manage to get permission to visit?"

It had been easy, strangely enough. I had faxed the company a letter explaining that I was a Chinese-American writer and chef. I mentioned that I had just passed the national cooking exam and wanted to "spread propaganda" about MSG to Americans. The next morning I got a call from Mr. Liu: "When would you like to come?"

In between the flavorful meals and the tours of the factory and town, I asked Mr. Liu to comment on the allegations of pollution. "We had a problem a few years ago, but it's all cleaned up now," he said breezily. "Anyway, we do so much economic good for the city. We've created jobs for thousands of people."

"Does Lotus Flower engage in any community service? Does the company contribute to philanthropic causes?" I asked.

"We give people jobs. Isn't that a community service?" he snapped, sounding very un-Communist.

The factory town's pollution was as pronounced as its new wealth. Plumes of smoke rose from a trio of smoke-stacks just beyond the company hotel. The air smelled like coal, and the sky was unnaturally opaque. I later met with Wang Jiaqin, the vice president of the China Fermentation Industry Association, who explained that since "the national government was becoming more concerned about the envi-ronment, factories have moved out of cities and into more rural areas." The Shanghai-based Heaven's Kitchen had switched to making fewer polluting substances, and rural factories like Lotus Flower took up the slack, Wang told me. (She didn't specifically name the company as an offender.)

Curious to find out more about the environmental damage, I called a local environmentalist, whose phone number had been passed to me through a friend. I had just finished an MSG-laden dinner with company representatives and had gone to my hotel room—a giant suite that a company official insisted was the only room suitable for a visiting journalist.

"It would be very inconvenient to meet you," the environ-mentalist said stiffly. He sounded distant, and the phone line crackled.

I explained that I was a guest of the factory and that I wanted to meet with him for a casual conversation.

"It would still be impossible," he said. "You can take a look at the river yourself. It's very close to your hotel."

"Well, how about we meet tomorrow, just briefly?" I suggested.

"You need official permission from the government to do an interview. And anyway, I'm no longer doing that kind of work. What I mean is, environmental protection will still be my work. It's still important. But we're not working on that project anymore."

I was confused; I hadn't mentioned any particular project. Clearly, I wasn't getting anywhere. "Okay. Well, how about I call you tomorrow?"

"Yes, why don't you do that?" A deep voice boomed on the other end. It wasn't the environmentalist. The line went dead.

I sat alone in the gigantic suite, waiting for the cops to come bursting into my room at any second. Meanwhile, and rather annoyingly, I couldn't stop feeling thirsty. I had eyed the water cooler in my room suspiciously when I first settled in, and now I was downing cup after cup of water from a cooler that proudly bore a Lotus Flower label. It must have been the MSG that made me so thirsty. Maybe it was also making me paranoid. As I guzzled my fifth cup of potentially carcinogenic water, I received a text-message reminder from my friend at the *Wall Street Journal*: "Don't drink the water!"

It was too late. I was going to be detained, deported, and, before long, I'd get cancer and die. My head throbbed, and I wondered if that was another MSG symptom. I got into bed and pulled the comforter tightly around my body, trying to ignore my paranoia and my thirst. I tossed and turned, convinced that I had identified yet another side effect: insomnia.

Maybe I didn't sleep well because I had to make multiple trips to the bathroom to pee. But it was more convenient to blame it on the MSG.

The next day, after the plant visit and another flavorful meal, one of the company representatives suggested I take a rest in my hotel room. I quickly agreed, but once I was alone, I headed to the river instead. The river hadn't been included in Lotus Flower's tour of the town, and I understood why when I arrived. A muddy dirt road separated the steep river-bank from clusters of shacks. Villagers squatted near the bank, holding up umbrellas on that drizzly day. They watched as fishermen cast their nets in the water, flinging them as far as they would go and pulling them back to shore. Black sludge dripped from the nets. For every three or four tries, the net trapped one or two fish that looked no bigger than sardines. Upstream, the Lotus Flower factory pumped out smoke. When I returned to Beijing, I would find out that MSG production releases ammonium nitrogen, a compound that is toxic in high concentrations. But even as I stood on the riverbank, it was clear enough that the Lotus Flower factory was not good for the environment.

Can you swim in the water? I asked a few of the locals.

"Of course," they said, as if it were a silly question. Of course they could swim. They could also fish, and they could bathe in it. They could do whatever they pleased. They didn't have any bad thoughts about Lotus Flower. "It's a very famous brand," said one of the locals proudly.

I tried to find a way to explain that the factory might be bad for their health, but I gave up. It would be like trying to convince Americans that MSG *wasn't* bad for their health.

Haihua, my escort, dropped me off at the train station that afternoon. When we said our goodbyes, she presented me with a heavy gift bag containing a new product line: Ginko Green MSG. I jostled my way into my train compartment with the case of MSG, but forgot to take it with me when I disembarked in Beijing. The next meal I ate at a restaurant tasted strangely bland. I asked the waitress if they used MSG. "Absolutely not," she said with a look of indignant disgust.

part two

NOODLE INTERN

6

SPLASHES OF BOILING WATER singed the back of my hand. Steam penetrated the pores of my face. I glanced uneasily into the gigantic wok and took a deep breath. With a noodle knife that looked like a giant razor blade, I was trying to grate the five-pound slab of dough I held on my left forearm into ribbons. *Daoxiao mian*—knife-grated noodles—were a specialty of Chef Zhang Aifeng's home province of Shanxi. When Zhang made them, they came out ridged and elegant, like party streamers. They slid into the wok with just a hint of a splash, like a succession of Olympic divers. Mine looked more like stretched-out wads of chewing gum, too fat and shaved at the wrong angle. They belly-flopped into the wok, like chubby kids at a community pool. Each splash of scalding water was an indictment from the noodle gods.

My cooking certificate, which I'd hoped would be my instant passport into Beijing's professional culinary world, had proved to be an instant gag instead, provoking howls of laughter from potential employers before they handed it back to me. My cooking school had a fairly good record at placing students in restaurant jobs, but a foreigner with a cooking certificate was another matter. Why on earth would a foreigner want to do something so lowly, something with the status of an auto mechanic?

Chairman Wang worked her way through the school's database of restaurants and came up empty-handed. "I'm sorry," she said. "Nobody believes that a foreigner will work in a kitchen." I thought perhaps the school's cooking teachers could accommodate me at one of the hotel restaurants where they worked, but government-owned hotels were off-limits too. Chef Gao balked when I asked if I could simply visit him at work. "Oh, no," he said. "We don't let *laowai* into our kitchen. Absolutely not." I had no better luck with the owner of a chain of Sichuan restaurants to whom a friend had introduced me. When I asked if I might visit one of her kitchens, she pretended to be preoccupied with mining her giant yellow purse for a vitamin to swallow down with her food.

So it was by the process of elimination that I ended up at Chef Zhang's noodle stall, in a humble canteen in southeastern Beijing. The area was full of furniture warehouses and home-improvement shops aimed at the growing middle class. Long tables with plastic chairs spanned the dining hall, and a series of stalls lined one wall. Zhang's booth was in the middle, underneath a sign that read REAL SHANXI SNACKS. The white-tiled kitchen was the size of a walk-in closet and was cluttered with a stove, a stainless steel counter, a sink, and an olive-green refrigerator that Zhang had purchased used for $30. It was the summer of 2006, a very hot time of year to be cooking in a windowless room.

Zhang didn't so much invite me to work for him as acquiesce; the day I poked my head in to ask if I could apprentice with him, he was frantically shaving a huge hunk of dough that he held up on a wooden board in front of him, like a violin. He was sweating so much that his white shirt looked practically transparent. The orders were coming in so fast that he had lost count of how many bowls of noodles he was supposed to make.

"Are the noodles ready yet?" a diner with a cigarette hanging from his mouth shouted from the mess hall. There were several more customers just like him, gruff and hungry. Two more middle-aged men arrived at the counter and impatiently demanded noodles. As the pressure mounted, I told Chef Zhang that I wanted to become a noodle chef; he didn't have the time or energy to say no.

He had big, expressive eyes and protruding ears that emphasized his perceptive nature. He had just turned forty-one, but his skin had a youthful glow, which seemed to come from the sweat of his labor as much as it did from his gung-ho outlook. A jade Buddha hung from his neck. He had a narrow waist, accentuated by the dirty white apron he tied tightly around his petite frame. But his lean, muscular arms, ridged with big veins, gave him a masculine air. If Chairman Wang had been my window into the lives of China's urban middle class, Chef Zhang was my introduction to an entirely different class of people, the struggling migrant workers with little time to complain about social ills or the graft of government officials. Most of them worked seven days a week for a meager salary, most of which they saved and sent back home to their rural families, in hopes of giving their children a better life than theirs.

Chef Zhang's noodle stall cost him $250 a month in rent. He had to sell about six hundred bowls of noodles a month to break even. His only full-time employee was his niece, who stood at the front of the kitchen, in a narrow galley with a counter, taking orders from customers and calling them back to him. Zhang called her Haizi, which wasn't her real name, but simply meant "child." Though she was twenty years old, she looked like a child, with a plump face and bushy bangs that flopped over her eyes. When the orders were ready, the customers, many of whom were furniture salespeople in the district, paid her and carried the food to one of the long tables. Chinese canteens were among the few places where you'd see Chinese eating alone; most other sit-down eateries were made for socializing as much as eating.

The canteen was staffed with two *ayi*, or "aunties," a euphemism for hired help. The aunties mopped the floors, cleared the tables, and helped when Zhang got too busy in the kitchen. Auntie Xu, who had a long face with a bored expression, constantly shifted her gaze around the canteen, reminding me of the busybody volunteers of the Party who did nothing but sit in the front courtyard of my apartment building, observing the comings and goings of everyone in the neighborhood. All over China there were people like that. I much preferred Auntie Feng, who had a cherubic face and a cheerful voice. "You're from America?" she asked me wide-eyed one morning not long after I started. "I have a distant relative who went there. She says your country is very clean. She tells me that you can wear the same white shirt for three days and it won't get dirty!"

The landlords, two brothers surnamed Han, spent their time loitering in the canteen, swatting flies, playing cards, and waiting for their rent money to come due. On occasion, they tried to get to the bottom of the mystery of why I was there.

"You know Chinese and English," said the older brother one day. "You seem to be educated. Why don't you get a job in an office building?"

But when the lunch rush began, my novelty waned. I was just another body serving up bowls of hot food.

An unofficial noodle-rice line runs across China, dividing the country the same way the Mason-Dixon Line once divided America. Since my parents followed southern Chinese traditions, I grew up on rice. Rarely did my mom cook noodles, and when she did, they were the Chinese-grocery variety of thin, dried egg or rice noodles, which she stir-fried or boiled in a hot pot.

In Beijing, giant bowls of chewy wheat noodles were a staple, as they were elsewhere in northern China — sometimes soupy, sometimes smothered in a thick gravy of tomatoes, meat, or soybeans. They were too filling for me; no matter how much I slurped and chewed, I could get through only half a bowl. But Beijingers like Chairman Wang and her husband complained that they couldn't get full on rice—only a bowl of noodles would do.

I had to admit: noodles were sexier. Rice was steady but boring—the varieties weren't so different and they went with everything, like a black T-shirt. Noodles came in endless shapes and sizes. For high-maintenance types there were "twelve-times hand-pulled noodles," which came out thinner than angel-hair pasta. For the less fussy there was *mian pian*— flat "noodle pieces" simply ripped from a long strand of dough. Cooking rice was easy—you put the grains into a rice cooker, added water, and waited for a light to tell you it was ready. But even simple noodles involved real work, the concerted effort of arms, wrists, and fingers to knead, twist, push, and pull, as beads of sweat dripped from your forehead.

The dough itself had endless permutations and combinations. White flour was the most common, but noodles could be made from virtually any grain that could be finely ground, including corn, buckwheat, and millet. They could be served hot or cold. Cold noodles — especially with sesame and peanut sauce — had a certain allure that cold rice would never have.

The first time I was exposed to the beauty of noodles was in China's northwest, when I was traveling on the Silk Road that once connected China with Rome. I was on an otherwise ordinary flight to Lanzhou, the stark and isolated capital of the dusty northern province of Gansu, a thousand miles west of Beijing. Lanzhou had the notoriety of being one of the most polluted cities in the world. As the plane began its descent, the flight attendant announced that we would soon be landing in "the home of the famous hand-pulled beef noodles."

That piqued my interest. Not long after I got off the plane, I climbed into a taxi and asked the driver to take me to the best hand-pulled beef noodle restaurant in the city. He deposited me in front of an eatery called Mazilu, a giant, high-ceilinged mess hall with long tables and benches. Though it was only two o'clock in the afternoon, the restaurant was about to close for the day, as the noodles they served were strictly for breakfast and lunch.

A woman at the register took my 3 *yuan* — 40 cents — and gave me a slip of paper and a pair of wooden chopsticks. In an open kitchen, a dozen men folded and pulled the dough, moving their hands up and down and side to side as if they were playing accordions. When the noodles reached the correct thinness, they were cut with a snap of the hand and tossed into a giant vat of boiling water. A quick bath later, they were drained and placed in a bowl of beef broth and topped with a dash of chili sauce and bits of beef.

I blew on the noodles impatiently, then risked a first bite amid the steam and burned my tongue. Nevertheless, I knew that I was in the presence of the best noodles I had ever eaten. They were thin and chewy; the broth was smooth and slightly spicy. I detected hints of peanut and sesame, which smoothed out the chili-and-coriander-flavored soup.

My noodle ecstasy continued farther west along the Silk Road. In the town of Xining, a few hundred miles west of Lanzhou, I devoured plates of *mian pian*, which a chef ripped to the size of large postage stamps and stir-fried with squash, tomatoes, onions, and mutton. The texture of *mian pian* was chewy but still al dente. The sauce, made of tomatoes, sugar, and vinegar, was at once sweet and sour. I left Xining with a pleasantly plump Buddha-like stomach, which proved to be fortuitous, since at my next stop, Tibet, I was awed much more by the temples and the people than by the staples of dried yak meat and ground barley paste.

So I respected noodles when I began working at Zhang's stall. But the idea of crafting the dough was as daunting as working in an unfamiliar neighborhood with an edgy, industrial feel. In cooking school I had grown used to the class routines and teachers; the noodle stall was gritty and real. In the humble canteen, cooking had consequences—Zhang's family depended on it, and I sensed that Zhang couldn't afford to make mistakes.

NORTHWEST-STYLE NOODLES (*LA TIAO ZI*)

1 pound Chinese wheat noodles
(preferably fresh, but dried will work)
2 tablespoons vegetable oil
1 medium yellow onion, diced
3 cloves garlic, minced
½ pound beef tenderloin, cut into thin slices
3 tomatoes, diced
1 cup coarsely chopped napa cabbage
1 green pepper, diced
½ cup ketchup (preferably Lee Kum Kee brand)
½ cup water
½ teaspoon salt
2 tablespoons Chinese black vinegar

Place a pot of water over high heat. When it comes to a boil, add the noodles and cook until tender (fresh wheat noodles will take about 3 minutes). Drain.

Add the oil to a wok and place it over high heat. When the oil is hot, add the onion and garlic and stir-fry for 1 minute. Add the beef and stir-fry until it browns, then add the tomatoes, cabbage, and green pepper and stir-fry for an additional 3 to 4 minutes. Add the ketchup, water, and salt and reduce the heat to medium. Once the liquid begins to boil, add the vinegar. Stir-fry for another 2 minutes, remove from the heat, and serve promptly over the noodles.

"So what's your restaurant going to look like?" Zhang asked me shortly after I had begun interning in his kitchen. He had opened the noodle stall just a few weeks before.

I was confused. Why did he think I was going to open a restaurant? On further discussion, it became apparent that he had assumed that the reason I wanted to work for him was so I could learn his secrets and put them to use in the restaurant he imagined I would open in America. He didn't mind. Why else would a foreigner want to labor in his kitchen?

The truth was, I was too much of a klutz to be a culinary spy. On one of my first days at work, I dropped and shattered a porcelain spoon. Then I lifted the counter and carelessly let it fold over on its hinge, catching it just in time to prevent it from smashing a pile of plates in its way. I was also consumed by my duties, especially after the lunch rush began. I was inundated with so many bowls of noodles that needed to be covered with sauce and garnished that I felt as overwhelmed as Lucille Ball in the candy factory. Despite the press, Zhang approached every bowl of noodles with the pride of an artisan. He didn't have the assembly-line fast-food mentality. He didn't seem aware that his customers were short on time and would have been almost as happy with instant ramen. Every so often customers canceled their orders and went to the Sichuan stall next door. I pointed this out to Zhang. With only the two of us in the kitchen, there were simply too many orders to fill.

"Can't you get a noodle-making machine?" I gently suggested one day. The machine would cost no more than $15, and I'd seen them in some restaurants.

"You can't use a machine to make noodles this good!" he huffed. "They won't taste the same."

By the time I left the shop, around three in the afternoon, I was so exhausted that all I had energy for was to get myself home, shower, and collapse into bed. I'd wake around six and think of Zhang. By that hour, he would have made his daily run to the nearby wholesale market and was probably washing vegetables and scrubbing the floor before closing for the day. He would go home soon, drift into a deep sleep, and wake up to repeat the same routine.

The noodle shop was open every day, including weekends and holidays. Zhang's routine was so repetitive, in fact, that he often forgot what day of the week it was.

"Is it Monday?" he'd ask on a Wednesday.

Until then, I'd associated such forgetfulness with carefree holidays when there was no set schedule, nothing in particular to do. It had never occurred to me that someone could lose track of time because there was too much of the same to do.

7

I LABORED IN THE BACK of a dingy noodle stall by day and dined in Beijing's poshest restaurants by night. I had recently become the food editor of *Time Out Beijing*, a monthly magazine that covered the capital's burgeoning art and culture scene. It was my first job as a professional food writer.

The magazine had two editorial teams, one Chinese and the other Anglophone, which published separate Chinese and English editions of the magazine. The government censored both, though I was surprised by how much the censors allowed when it came to expletives and topics like sex and the gay community. I wrote for the English version, which was bolder than the Chinese edition and attempted to publish something virtually unheard-of in China: the independent review.

Beijing restaurateurs, by 2005, were hatching ambitious projects that were a far cry from the blandly decorated state-owned restaurants that once dominated the city. Some were successes, like one Peking duck house that served sumptuous roasted birds and novel side dishes that didn't lose sight of their Chinese origins. Others went too far with their avant-garde gimmicks. When the second location of a well-known Beijing eatery opened, I panned a discordant dish of wasabi, green tea,

and mango prawns. The décor was no better; it was "no surprise to find out that one of the first few events to be held at
the new Green T. Living was a memorial service," I wrote with
a critic's glee. "The place, which resembles a mausoleum, is a
fantastic setting for a funeral . . . The restaurant is kept at an
uncomfortably cold temperature, not unlike a tomb."

I winced when I saw the published copy; I hadn't quite realized when I wrote the review that it would sound so brutal
in print. It sent restaurateurs and foodies atwitter in a city
unused to candid reviews. One reader responded, "That place
has been serving crap food in pretentious surroundings with
shitty service for years now and yours is the first publication
to point out that the Empress"—a reference to the precious
Chinese woman who owned the restaurant—"has no clothes."

Locals found my review methods peculiar, to say the least.
I showed up in dining rooms unannounced, as it was done in
the United States and Britain. If I had to make a reservation,
I did so simply under "Lin." At the end of a meal, I paid the
bill in cash, and I was later reimbursed by *Time Out.* "How interesting," said a Chinese friend when I described this. "So you
go to the restaurant as if you were an ordinary diner." Sucking in her breath, she said, "That's not how it's done in China."

Chinese journalists always announced their arrival; they
never had to pay for a meal. If the food was bad, they wrote
about the ambiance. (Usually, the worse the food, the better
the ambiance, one journalist told me cheerfully.) After enjoying the free meal, they often received a *hong bao,* a "gift" tucked
in a red envelope, to ensure the review would be positive. To
further hedge their bets, restaurateurs made friends with pub-
lishers and editors (by inviting them to meals, of course) so
they could edit reviews of their own establishments before pub-

lication. Sometimes the arrangement was even more overt: reviews appeared that were no more than paid advertisements, though they were never acknowledged as such in print.

With little freedom of the press, publications didn't have much of a moral compass. People valued *guanxi,* or connections, more than they valued candor. To journalists and restaurateurs it was a cozy arrangement; the journalists were well fed and the owners did a brisk business.

One afternoon, upon leaving a restaurant opening, I was shocked when I opened the press kit—put together by an American PR firm, no less—to find a *hong bao* of $25 tucked inside. More plugged-in journalist friends were unimpressed. They informed me that they had received bribes of $100 or more. I was still green.

More shocking than the *hong bao* were the exotic dishes Chinese liked to eat. I had abandoned my pastime of asking my Chinese friends to name the most unusual thing they had ever eaten after a friend recalled the time her mother had served her a stir-fried dish containing an unidentifiable meat. "It's lamb," her mother had assured her, waiting until she had finished to confess that it was a human placenta, which the Chinese (and other cultures, including the French) have traditionally valued for its high nutritional content. As an American, I was easily shocked by such stories, but beneath the cringe factor, there was a beauty in the economy of Chinese attitudes toward food. Nothing was taken for granted, and nothing went to waste. In the not-so-distant past, a finicky eater would have starved. And if you were a meat eater, where were you supposed to draw the line anyway?

My new position began to put my attitudes to the test.

During the Year of the Dog, the editors of *Time Out* thought it would be fun for me to write about eating dog stew. More than a hundred restaurants in Beijing served dog meat, although animal rights advocates were urging a ban in time for the 2008 Olympics, when China would be in the international spotlight.

I had invited my friend Xin, the placenta eater, to join me. She was waiting for me in front of the restaurant, which appeared to be doing a bustling business. When we sat down and were presented with a simmering stew, Xin immediately dove in. I hesitated, trying to psych myself up for the assignment.

"Come on, you've already gone this far," Xin said, as if she were a mother nagging a child. Dog, she reminded me, was supposed to improve one's circulation. After downing a glass of beer, I summoned the nerve to pick up my bowl, which Xin had filled with meat and broth. I took a sip. It was hearty and smooth. I dipped a piece of meat into an accompanying bowl of chili sauce. The meat looked—and tasted—like stringy lamb.

Xin pressed the analogy. Why did I find it acceptable to eat the flesh of a bleating baby sheep but not a scrappy mutt from the countryside?

"Well, most Americans have a special bond with dogs. We keep them as pets," I said. Chinese were also becoming avid dog owners, I pointed out. Weren't they bothered?

"But it's not like you're eating *your* dog," she said.

Would she mind eating her neighbor's dog? I almost asked, but I supposed she had a point. After all, I had met bird owners who ate chicken. (Incidentally, in the course of my eating adventures, I determined that the common stereotype that anything exotic "tastes like chicken" is simply not true.) I hit my limit, however, after the manager informed me that the dogs were butchered when they were four months old, so the meat was still tender. I was eating puppy. On the way home, I was struck by a migraine: karmic retribution.

A few months later, while I was filming a television segment for CBS News on Chinese street food, someone coaxed me into eating a fried scorpion on a stick. But a harsher challenge lay in wait: I was to review a restaurant called Strong in the Pot, which specialized in the genitalia of male animals. Chinese believed that eating these private parts increased a man's virility. A common saying in Mandarin was "*Chi shenme, bu shenme*" — eat what you need to repair.

The first hurdle was finding a dining companion, which had never been a problem before. "I'll be out of town," said some invitees, disingenuously, while the more candid responded with a reflexive "That's disgusting!" Only Jimmy, a Chinese friend of mine, was brave enough. A married public relations executive, he confessed to needing help in the love department: "Men are so stressed out at work that it's hard for them to perform at home," he opined. "When can I pick you up?"

The restaurant was just a block — albeit a very long Beijing block — from my apartment, but I had never noticed it before. It was an inconspicuous storefront with shaded windows that gave no indication of the activity inside, like an adult movie house. Beyond the small lobby was a hallway that led to a row of private rooms. A woman in a black pinstriped suit led us to one of them, which contained a round table with a lazy Susan. I was grateful for the privacy, which seemed to relax everyone, ever so slightly.

At the last minute, I had persuaded two more friends to join us. Victor, a Chinese Ph.D. student at Peking University, was a curious bystander; the idea of eating an animal penis turned his stomach the same way it did mine. Jon, an American who had been living in China for eight years and was married to a Chinese friend, was simply adventurous. "After living here for so long, I figure it's about time I try it out," he said.

The manager introduced the woman in the pinstriped suit as Miss Zhao. With highlighted hair pulled back in a ponytail, she looked barely of age, although the suit lent her an air of professionalism.

"Where's the nutritionist?" Jimmy asked. When I had booked a table, the manager had told me that a nutritionist would guide us through the meal.

"Miss Zhao *is* the nutritionist," said the manager.

"Are you certified?" I asked.

"She is, by our standards," the manager said, before Miss Zhao could reply.

Miss Zhao had a short speech prepared. "Since ancient times, people have worshiped *bian*," she said, using the proper term for penis. It wasn't only men who would benefit from eating *bian*, she asserted; it was good for everyone's complexion. I wondered if the restaurant's fare had helped Miss Zhao maintain her silky skin. I was impressed by the range of meats on the menu, which included horse, deer, sheep, and snake reproductive organs. Miss Zhao brought out a Canadian seal *bian* for us to see. For $400, you got a small, shriveled appendage, accompanied by a pair of testicles.

Miss Zhao noted that *bian* was, in the lingo of Chinese medicine, a "warm" food, meaning that it was best eaten during the winter. Eating too much could increase one's *huoqi*—internal heat—sometimes to the point of inducing a nosebleed. "I'd say it's probably a bad idea to eat *bian* more than once a week," she said.

I dutifully wrote everything in my notebook, dreading the

moment I would have to put down my pen and pick up my chopsticks.

It was comforting yet disturbing when the first dish of bull *bian* arrived looking like an ordinary Chinese appetizer. The *bian* was cut into long, thin slivers that might have come from any other part of the bull. Tossed with shredded onions, bell peppers, and vinegar, it was a little tough but palatable.

The food and the conversation went steadily downhill. The next course was several kinds of *bian* boiled in broth, along with a turtle, beef bones, deer antler, and ginseng. (Exotic foods were often cooked in soups, which were believed to have restorative powers.) The wide range of ingredients didn't improve the dish, and the spices couldn't mask a medicinal taste. To me, it seemed that such soups were just an expensive way for Chinese to impress their dinner guests. The meats varied from chewy to extremely rubbery, though I concluded that if I were ever forced to eat *bian* again, I'd opt for donkey, which seemed to be the most mild and tender of the lot.

Just as I was about to bite into the sheep testicles, Miss Zhao shrieked, "They're only for men! If you eat the balls, you'll grow a mustache."

Victor, spooked by the dishes, had left after a single bite of the shredded bull. But Jimmy and Jon actually seemed to be enjoying themselves. They continued to eat long after I had stopped, washing down the food with numerous beers.

"Is it true that foreigners have bigger *bian* than Chinese?" I heard Jimmy ask Jon. Whether it was the beer or the cuisine that had emboldened him, I wasn't quite sure.

8

I FIRST MET CHEF ZHANG about half a year before I started working for him. He was running a noodle shop not far from the canteen where I apprenticed. I was struck by how he said goodbye to a couple who had finished eating. "If you have any suggestions, don't hesitate to tell me," he said as he walked them to the door. In a country where people often tried to cut corners, it was refreshing to hear someone who cared about making good food and pleasing customers.

Zhang came from Shanxi, the poor, dusty province where Chairman Wang had been sent during the Cultural Revolution. He was born into a family with five sons and one daughter. Zhang's parents were too poor to raise all six children, so they decided to give two of them away. Zhang, as son number four (an unlucky number in Chinese tradition), and his sister (unwanted because of her gender) were sent to live with a childless aunt and uncle twenty miles away. Zhang was a toddler at the time, and his sister was twelve.

"They didn't know how to read," Zhang said. "They were poor farmers who raised sheep." His adoptive mother died when he was five years old. A few years later, his sister married and left home. From then on, it was just Zhang and his uncle. "I considered him my real father," Zhang said. "I had a very strong bond with him."

Zhang was still a child during the Cultural Revolution, but he remembered the communal kitchens the government instituted. "If local officials saw smoke coming from your chimney, you'd be in big trouble. They'd take away all your cooking utensils. They'd make you wear a sign around your neck and gather everyone in the town to denounce you."

When the Cultural Revolution subsided, standards of living improved in the countryside. The government distributed food to each household, and because he and his uncle made up one household, they always had enough to eat. "The government gave us a small amount of meat each month," Zhang recalled. *Daoxiao mian*, the sliced noodles that later became his livelihood, figured prominently in their diet. "But rather than wheat flour, we grew millet and buckwheat and used them to make noodles."

By the time Zhang reached his teens, his uncle was in his sixties and had health problems, so Zhang took over most of the farm chores. Besides going to school, he raised pigs, cooked meals, and looked after the sheep. "I had to pick herbs in the mountains to pay for my education. Growing up, I didn't have anything. I had to do everything myself. I knew I couldn't stay at home waiting for something good to happen to me."

With all his responsibilities, he lagged in his studies, and didn't complete junior high until he was seventeen. He wanted to go on to high school, but his uncle didn't have the money. (Though technically free, schools levied numerous fees on parents that padded teachers' and administrators' incomes.) In any case, the uncle considered education a waste, especially since he needed Zhang to farm.

But China had embarked upon its massive transformation from an agrarian to an industrial society. The economic reforms of the early 1980s had begun, and the nation was opening up to the world. Zhang saw an opportunity to break free of old traditions. A big coking factory had opened near his village, which converted coal to coke, used as a fuel and to make steel. But the workers who were hired all had backdoor connections. Zhang decided to write a letter to the factory's Communist Party secretary, describing his poverty-stricken boyhood and his aspirations to rise above his background. The official, moved by the letter, hired Zhang as an office assistant.

Though the job wasn't much, it was a big opportunity for a kid who had been raised by illiterate farmers. His salary was $5 a month, enough so that he could think about settling down. He courted one of his coworkers and married her when he was twenty-four. A year later, she gave birth to a daughter. The government had begun enforcing its single-child policy, but the rules were lenient in the countryside. Rural families whose first child was a girl could have another child after waiting four years. When his daughter was four, his wife gave birth to a son. Unlike many rural people, Zhang didn't value boys more than girls. "But I did it to make my father happy," he said, referring to the uncle who had loved and cared for him. The birth was bittersweet; soon after his son was born, his uncle died.

Zhang had risen in the company, and by the mid-1990s he was making $60 a month as a midlevel manager. With the burden of caring for his guardian now lifted, he decided that he wanted to go elsewhere to pursue his career. Beijing was the most obvious place, as the largest city in the region and only an overnight train ride away. Zhang had an aunt there who could help him find a job. His wife stayed at home with their daughter and newborn son.

Zhang arrived in Beijing in 1997. The possibilities for employment were limited. Many migrants went into construction work or other jobs involving manual labor, but Zhang aimed to do something more skilled. His aunt found him a job in a Sichuanese restaurant. He had never worked in a restaurant before; he'd chosen the job because, at thirty, he felt that he was too old to learn much else. "Everyone has to eat, so I figured I would always be able to find a job," he said.

He washed vegetables for several months, until he was hired as a cook at Yushan, a successful restaurant where he would spend the next eight years. He started with a salary equal to his pay at the coking factory.

"I had to learn everything, so I couldn't ask for more pay," he said. "But I still thought my prospects would be better." The job provided free room and board at least, which meant that Zhang could send home nearly his entire paycheck each month. "I kept a few dollars for myself to buy cigarettes," he recalled.

He learned that restaurant cooking was nothing like Shanxi home cooking. "When I cooked at home, it was very simple. We ate what we needed to get full. Maybe some noodles or steamed bread with some pickled vegetables. But in Beijing I realized that flavor was important," he said. The restaurant introduced him to oyster sauce, Sichuan peppercorns, lemons, and Chinese barbecue sauce. "I had to learn the names of all the vegetables and all the dishes. The hardest part was learning how to use the wok. I didn't know how to pickle anything. I didn't know how to *guoyou*," he said, referring to a method of flash-frying meat to seal in the juices.

Since Zhang had started his cooking career so late, he apprenticed under a master who was the same age. On the job, he learned the skills I had learned at the cooking school: how to chop, season, and wok-fry. Zhang moved up the ranks at Yushan until he became a senior chef, with a salary of $200 a month. His schedule didn't vary: he woke at 9 A.M. and began prepping for the lunch rush an hour later. After a two-hour break in the afternoon, it was back to the hot, steamy kitchen until 9 P.M. He had only two or three days off a month, and he rarely ventured out of the restaurant. While his colleagues gambled, drank, and watched television after work, he often sat by himself in a corner of the dining room, reading novels and writing in his diary until the early morning hours.

Though I didn't know Zhang at the time, I had been to Yushan while he was still working there. I had become friends with one of the waitresses, and I visited her occasionally. It was the type of place where noise—the din of conversations, the clink of chopsticks against porcelain, the shouting at waitresses— bounced off the walls and echoed across the room, creating *renao,* a chaotic vibe that Chinese enjoyed. Yushan was one of the vestiges of the old state-run system, still under government control, unlike many other restaurants that had converted to private enterprises. The restaurant's location, next to the Temple of Heaven, made it a convenient stop for tourists. The Communist Party milked its connections with the tourist bureau to bring in visitors by the busload. A gift shop near the second-floor lobby sold knickknacks, like silk-embroidered cell-phone covers and stuffed toy pandas.

The menu also catered to international travelers. Unlike most restaurants in Beijing, which specialized in the cuisine of a particular region, Yushan tried to cover all the bases. Chefs carved roasted Peking ducks, and waitresses served up bowls of *mapo* tofu. Guests dined on dim sum–like snacks that were once served to Qing Dynasty royalty and had become mainstays of tourist-trap eateries. A few exotic dishes also graced the menu: camel's feet, at $15, was the most expensive of these items.

Qin, my waitress friend, whom I had met while eating at the restaurant, came to Beijing in 2003 with a dozen classmates from a vocational school in Sichuan. They had specialized in hospitality, and their program set up an internship after they finished their studies. The internship proved to be nothing more than a waitressing job at Yushan. The students' parents, many of whom were farmers, were happy to send their children to the big city. The job meant one less mouth to feed and money wired regularly back home. After the school deducted a monthly internship fee of $6, the waitresses made about $100 a month, depending on their bosses' generosity. If business was good, they received a little extra, but in wintertime, when tourism fell off, their pay might be docked by as much as 10 percent. The girls were supposed to get two days off a month, but during major holidays, they often worked several weeks without a break.

The young women lived in a narrow, dreary, fluorescent-lit alcove attached to the restaurant's kitchen. Each woman was assigned to one of the bunk beds pushed up against the walls. The only door led through the kitchen. If a fire had broken out in the middle of the night, there was a good chance the girls would have died. Despite the dismal setting, the dorm turned into a slumber party after the evening shift ended: they bounced on one another's beds, sang Mandarin pop tunes, and snacked on Lay's potato chips.

By the time I met her, Qin was eighteen years old and had been living in Beijing for two years. Although I knew that she and the other waitresses supported themselves and helped out their families, I found it difficult to think of them as adults. They seemed so innocent and sheltered. Qin was the youngest, and in the evenings, she changed out of her waitressing outfit, a red and gold embroidered dress, and scrubbed off her make-up. With her hair in a ponytail gathered above one ear, wearing jeans and a hooded white sweatshirt, she looked like a student you'd find in a schoolyard, not the back of a restaurant. In seniority-obsessed China, her coworkers called her Meimei, "little sister." Although she was only five feet tall, she carried herself with remarkable poise and confidence.

"Some people call Sichuanese 'rats' because we're very short, very clever, and sometimes very cunning," she told me once with an impish grin.

One afternoon, I found Qin in a serious mood. "Before I came to Beijing, I couldn't have imagined anywhere more spectacular," she recalled. "I had heard so much about the history and the culture. I envisioned Mao Zedong and all the crowds at Tian'anmen Square. It was the best place in the universe. It was developed in a way that Sichuan wasn't. But when I went back to Sichuan this year, I didn't want to come back. I noticed that things were getting better in Sichuan, too."

Then she switched back into girlish mode and belted out a line from a famous Chinese song: "The Great *Waaaall* is *looonnng!*"

Every morning, the manager took attendance to make sure the girls were there. Qin made sure her fingernails were neatly trimmed, lest she be fined. She and a couple of other waitresses had their pay docked for getting their ears pierced, though no one had told them that earrings were prohibited. When a Japanese customer left her a $5 tip, however, she knew the rule: she gave half of it to her boss.

"Japanese are very polite, very kind," Qin said. That had surprised her. In school, she had been taught that the Japanese were cruel. Her textbooks had described how they had slaughtered an estimated 300,000 Chinese in the Nanjing Massacre of 1937, and many Chinese held negative opinions of Japanese. The waitresses had spent so little time outside the restaurant that their worldview was shaped by foreigners who ate there. "Americans like anything that's sweet and sour," she told me.

She preferred waiting on foreign customers. Chinese didn't treat waitresses with much respect. One word for waitress — *xiaojie* — was a euphemism for "prostitute." The increasingly common title for wait staff was the gender-neutral *fuwu yuan*. But the term, which literally means "helper-person," didn't seem much more polite than *xiaojie*. Chinese diners ordered the *fuwu yuan* around like servants, rarely saying please or thank you.

Qin's best friends at the restaurant were the Cai triplets. The boss was kind to them because he thought that the identical trio of hostesses, with their long wavy hair and dimpled smiles, increased sales. "Back in Sichuan, we were the most popular girls in our grade," Qin said. "We were inseparable. We ate all of our meals together."

Everyone called the eldest of the triplets Laoda, which meant "old-big." It struck me as funny, because she was neither old nor big. But she was curious. She constantly asked me questions: How much was my rent? How much did I make every month? She told me that her parents, who had already given birth to one child, had been fined triple the amount for the second delivery of the three girls. The fine had been a big burden because her parents were poor farmers. "What are farms like in the United States?" she asked.

I was at a loss to explain; the last time I had been on an American farm was on a first-grade field trip.

Qin and her coworkers had recently bought cell phones, and they eagerly added my number to their directories. Scrolling through Laoda's saved numbers one afternoon, I discovered that she had five hundred entries. She had recorded every phone number she had ever come across, including those she had seen on billboards and in newspapers.

"Just in case," she said earnestly. "It's more convenient that way, in case I need to make a phone call."

Every so often I would receive a text message from Qin or one of the triplets. "When are you coming to visit us again?" they asked. Now and then I dropped in unannounced after the dinner shift, and never failed to find them there. None of the waitresses went out after work. They showered in the kitchen, behind a sheet. Sometimes they watched television on a small black-and-white set that was perched next to the night watchman. And their beds were always inviting, since they had been on their feet all day.

One blustery March afternoon, I received a text message from Qin saying that she had changed her phone number. I didn't think much of it because Chinese weren't loyal phone customers; they constantly changed their service plans to get better rates.

I sent her a message telling her that I planned to visit the restaurant that night.

She wrote back to tell me that she wouldn't be at the restaurant.

Not at the restaurant? I wondered. Where else would she be?

She informed me that she had quit her job and moved out of the alcove. She was going to help a friend of hers who had just opened a noodle shop. We agreed to meet near the shop the next day.

When I spotted Qin's tiny frame amid a swarm of shoppers just outside a new superstore, the hood of her white down coat was pulled over her head, and she was jumping up and down to stay warm.

Ying, another Yushan waitress, accompanied her. She too had quit, Qin explained as we headed to her friend's noodle shop, down a nearby alley.

"Every month, our salary was getting smaller. Even when the restaurant was doing well, we were only getting sixty to seventy dollars. It was too little. We couldn't live on it," Qin said. The salary cuts had led to an exodus of staff.

Qin slipped into waitress mode as soon as we entered the restaurant. "Here, sit down," she said. "Put your bag down. What would you like to eat?" She poured me a steaming cup of tea. A couple entered the shop, and Qin repeated the routine just as smoothly. She wasn't wearing her red and gold embroidered dress anymore, just blue jeans and a sweater; nevertheless, she was a polished waitress.

Though Qin had said in her text message that she was starting work in the noodle shop, she now acknowledged that the friend who ran it probably couldn't afford to pay her even a modest salary. She wasn't sure what she was going to do or where she was going to go. It was her first day on her own in the real world, yet she didn't seem nervous.

While Qin busied herself with customers, Ying stood in the

kitchen doorway, embracing a pimply-faced man with spiky brown hair. I knew the waitresses were old enough to have boyfriends, but I was startled because it was the first time I had seen any of them being affectionate with a member of the opposite sex.

The noodle shop was essentially a shack. The corrugated plastic roof and walls and the sliding front door didn't keep the place very warm. But it was clear that the owner had done his best to create a comfortable atmosphere. The eating area held six clean tables with stools, and a television in the corner blared the same sappy soap operas broadcast in living rooms across China. A giant laminated menu on the wall read "Specially Created Noodles in a Pot," and the shop was spotless.

After walking a couple to the door, Chef Zhang, the shop's owner and later my noodle master, placed a steaming pot of noodles in front of me, gently, as if it were made of fine china rather than rough clay. The noodles were cooked with pork ribs, seaweed, and mushrooms, and they were a welcome respite from the windy afternoon. Though it was the tail end of winter, everything in the bowl tasted as if it were fresh from a summer market. With a touch of vinegar and chili oil, I slurped down the entire bowl. At 80 cents, it was twice as expensive as a typical bowl of noodles, but Chef Zhang was betting that a rising middle class would be willing to pay more for quality.

As Zhang busily worked away in the kitchen, Qin told me the basic details of his life. He was from Shanxi, and he too had worked at Yushan until a few weeks ago. This was the first restaurant he owned. Though he didn't have the money to hire waitresses, she and Ying were happy to help him. He had been the girls' favorite chef. He acted like a counselor to them, happy to listen or give advice to anyone who approached him after work, as he sat in a corner sipping tea and reading. He didn't like to drink or play mahjong.

While I sat slurping his noodles, the chef wrote me a note that briefly explained his plan to open a chain of noodle shops and asked for my help in publicizing his venture. Qin had told him that I was a food editor at a magazine. I told him I'd try my best to help him, though I knew his shop was too far from the usual expat hangouts, geographically and otherwise, to be featured in *Time Out*.

Zhang had persuaded a twenty-eight-year-old colleague to go into business with him. He had a total of $5,000 from his savings and what he'd been able to borrow from several friends. He had moved into a friend's home, where he could live rent-free. He was negotiating unfamiliar streets by bicycle to get to the shop, and was unsure if he would make enough to cover his expenses. After ten years in a factory and eight years in a state-owned restaurant, Zhang was finally his own boss.

When I finished my noodles and tried to pay, Zhang refused my money. I figured he was just trying to be polite. I placed a 10-*yuan* note (about $1.20) on the table, and he picked it up and put it into my bag. When I again tried to offer him the money, he stood in front of me and puffed out his chest to block my way. What seemed to start as a courteous gesture turned tense, almost hostile. It wasn't about hospitality; it was a matter of pride.

"Don't worry. Don't be polite," he said with a frown. "Don't take pity on me because I'm from the countryside."

The shop where I met Zhang survived for two months. The next time I called him, he had moved his operation a few miles away, to the canteen where I joined him as an intern. The rent was cheaper there. The canteen served the employees of a shiny new furniture emporium where wealthy Beijingers came to browse an assortment of decorative fountains, palm trees, and sofa sets. The employees wanted a quick and cheap lunch, so Zhang simplified his menu and lowered his prices.

Zhang told me that the previous shop hadn't lasted because his partner had abruptly decided to get married and move back to his village. One morning, after we had worked together for a while, he said, "I'm going to tell you something that's going to make me lose face." He bowed his head and said, "Our first shop didn't work out because my partner was sloppy. He didn't grate the noodles well. He didn't pay attention to the vegetables either. Once he served rotting ones to our customers." Zhang and his partner fought over this, and Zhang decided he couldn't work with someone so careless.

On most mornings, Zhang would light a stick of incense and set it in front of a small gold-colored Buddha hidden under the counter. The smoke drifted lazily for hours. An offering of three apples sat in front of the statuette. Zhang became a Buddhist around the time he quit his job at Yushan. Though Buddhism was the traditional religion of many Chinese, it had been outlawed after the Communists came to power. Only in the past two decades had the government started allowing citizens to practice Buddhism again, along with four other religions: Taoism, Islam, Protestantism, and Catholicism (though the Vatican had yet to recognize the Communist government as the rightful ruler of China). The regime still closely monitored churches and temples.

Zhang went to a temple near Yushan that was damaged in the Cultural Revolution but had since been restored. He told the monks there that he wanted to officially become a Buddhist. The monks performed a ceremony in which he burned incense while prostrating himself before a statue of the Buddha. When the ceremony was over, he donated $12 to the temple, and the monks gave him a booklet certifying him as a bona fide Buddhist.

Zhang thought that adhering to a blend of religion and superstition could help him succeed. Being Buddhist was good for business—at least it didn't hurt to ask the Buddha for luck. While negotiating with his landlord and suppliers, he used a different name, Zhang Miao. His real name, Zhang Aifeng, was unlucky because its character strokes lacked the symbol for water, one of the five main elements.

Zhang added that his religion had brought a sense of calm. I could see it in his dealings with Haizi, his niece and only employee. She looked like a sweet youngster, but she had a mischievous streak. In between taking orders, she sneaked bites of the food Zhang was making for his customers. She took a perverse joy in dumping the leftovers that Zhang planned to eat himself. She was the hyperactive type, and just watching her exhausted me. One morning, she let out a bloodcurdling scream as she flapped a newspaper in her hand.

Zhang and I looked up from the kitchen. "What?" he said.

"Look!" she gasped, running into the kitchen, pointing to a picture. It was a photograph of Siamese twins. "I've never seen a two-headed baby!"

The most exasperating thing was that she could not master her job. The orders she fired back to Zhang and me were often bewildering.

"Three *daoxiao mian!* One vegetarian! No parsley!"

Did that mean three or four bowls total? No parsley on any of them? We had no idea. Nothing was written down; the orders came too fast. Zhang and I tried in vain to keep track of them while balancing steaming-hot bowls in our hands. Several times a day, customers sent back noodles they hadn't ordered.

When this happened, I shot Haizi a look of annoyance, but Zhang usually forgave her and returned to the stove to make a new order. Once in a while, though, even he would lose his Zen-like cool. "HAIIIIIZZZZZIIIII!" he'd scream. Unfortunately, the whirring of the huge fan above the stove drowned out his voice. Haizi continued to relay orders and shove money into her apron pocket.

In the first few days of my internship, Zhang was hesitant to let me touch the flour. "You might not have the strength to knead," he warned as we prepped in the kitchen one morning. Haizi, poking her head in from the front, put it more simply: "It's a man's job."

The flour, in a paper sack slumped against the kitchen wall, remained off-limits to me. I figured that after the carelessness of his former partner, he was slow to trust anyone else with the job of kneading and grating. So I learned how to do everything else. After I had chopped the garnishes, I washed the dishes. When the dishes were done, I moved on to the eggs. Zhang started the day by boiling several dozen eggs with tea leaves in a large pot. The shells darkened to the color of molasses as they boiled. The white inside turned beige and had a smoky flavor, with none of the sulfuric scent of hard-boiled eggs. It made the perfect portable breakfast, and customers who wandered in before lunchtime often purchased a couple of tea eggs to tide them over.

I learned how to make appetizers. Unlike the appetizers in

a typical Chinese restaurant in the States, deep-fried and served with heavy dipping sauces, Zhang's appetizers were healthful, simple, and tasty. *Pai huanggua*—smashed cucumbers —was one of my favorites. The name of the dish essentially explained the method. Cucumbers in China are spindly, with prickly bumps on them. Zhang cut them into inch-long segments and whacked them with the side of his cleaver so they broke into little pieces. He threw them in a bowl and tossed them with garlic, soybean oil, sesame oil, and vinegar from his native Shanxi.

The same flavorings went into Zhang's *liangban doufu si*, or cold-tossed shredded tofu. The long strands of dried tofu had a springy, chewy texture similar to al dente noodles. To give the dish an extra boost, he sautéed Sichuan peppercorns, garlic, and chili peppers in cooking oil to infuse it with their flavors. When they were on the verge of burning, he removed the wok from the heat and strained the oil through a sieve, leaving it with a scent I could have bottled and sold as perfume.

TEA-INFUSED EGGS (*CHAYE DAN*)

6 large eggs
¼ teaspoon salt
¼ teaspoon powdered chicken bouillon
4-5 cloves
3 star anise
1 leek, white part only, cut into 1-inch pieces
2 thin slices ginger
2 teaspoons loose jasmine tea leaves

Fill a pot halfway with cold water and add the eggs. Bring to a boil and cook over medium heat for five minutes. Remove the eggs from the pot and tap them gently against a hard surface, cracking the shells but leaving them intact. Return the eggs to the pot and add the salt, bouillon, cloves, star anise, leek, ginger, and tea. Cook over medium heat for 20 to 25 minutes. Remove from the heat and allow the eggs to steep in the pot for at least 15 minutes before serving.

SMASHED CUCUMBERS (*PAI HUANGGUA*)

2 cucumbers
4 cloves garlic, minced
1 tablespoon sesame oil
2 teaspoons Chinese black vinegar
¼ teaspoon salt

Slice the cucumbers crosswise into 2-inch pieces. Using the flat side of a cleaver, whack the cucumber pieces once or twice so they break into smaller pieces. Transfer them to a bowl and toss with garlic, sesame oil, vinegar, and salt. Marinate for at least 15 minutes or up to several hours before serving.

COLD-TOSSED SHREDDED TOFU
(*LIANGBAN DOUFU SI*)

¼ cup vegetable oil
4 whole dried red chili peppers
1 leek, white part only, cut into 1-inch pieces
2 thin slices ginger, thumb-sized
4 cloves garlic, 2 left whole and 2 minced
5 whole dried Sichuan peppercorns
¾ pound shredded tofu (*doufu si* or *gansi*, available at some Chinese markets)
½ green pepper, diced
½ medium carrot, diced
1 tablespoon coarsely chopped parsley leaves
¼ teaspoon salt
¼ teaspoon ground white pepper
1 teaspoon Chinese black vinegar
1 teaspoon sesame oil

Pour the vegetable oil in a wok and place over high heat for 2 to 3 minutes, then add the dried chili peppers, leek, ginger, 2 cloves of garlic, and peppercorns. When the oil is very hot but not smoking, remove the wok from the heat and strain the oil, discarding the seasonings.

In a large bowl, toss the tofu, green pepper, and carrot. Add the minced garlic and parsley. Pour the seasoned oil over the tofu mixture and toss. Add the salt, pepper, vinegar, and sesame oil, toss, and serve immediately.

The dilemma that plagued me the most about my internship was how to get there. Beijing was a gigantic city, as sprawling as Los Angeles but with a population four times as large. I lived in the center of town, and Zhang's shop was in the far southeast corner. By subway and bus, the trip cost 60 cents and took one hour. By taxi, it cost $4 and took twenty minutes. I could afford the taxi, but I felt guilty showing up in one at the noodle shop, where Zhang's total profit in a day was often less than the taxi fare; the canteen aunties made $2.50 a day.

In the mornings as I lay in bed in a foggy state, summoning the energy to turn off my beeping alarm, I tried to psych myself up for the overcrowded subway and bus. But because I usually ran late, I ended up hopping into a cab. Embarrassed by my lavish habit, I would make the driver pull over down the street from the shop and walk the rest of the way.

"Wow," Auntie Feng would say, shaking her head. "Coming all the way from Dongzhimen. How long did the bus ride take? You're such a dedicated worker." I would just nod and hang my head. My shame was interpreted as modesty, which further boosted my status in the canteen.

Finally I came up with a compromise. Mornings, I took a taxi to work. By a stroke of luck, while most of Beijing traffic was at a standstill, the route from my apartment to the noodle shop was unclogged. In the afternoons, I boarded a lumbering bus. Half an hour later, I transferred to the subway, changing trains midway before it deposited me at a station that was a ten-minute walk to my house.

With taxis so cheap, I had previously spent very little time on public transportation. Taking the bus and subway exposed me to a side of the city I had missed. I found out that buses were a means of conveyance not just for people but also for large objects. When the bus doors opened at one stop, half a dozen workers clambered on, each hauling a fifty-pound sack of flour. Another afternoon, a man in the center of the back row steadied a ten-foot-long billboard that stretched into the aisle in front of him. The billboard featured smiling models squeezing eye drops into their eyes. When the bus stopped, he swung the billboard like a door to let people in and out of their seats.

Men in shabby suits who appeared to be from the countryside stared out the windows in awe of all the cranes and skyscrapers. At the subway station, a man begged for money. He was so horribly disfigured that I looked away, and then could picture only his swollen red eyes when I tried to recall his face.

One afternoon, as I was walking from the bus stop to the subway station, I spotted someone I knew crossing the street. I cringed at the possibility that I'd have to stop and chat, because my past encounters with her had been awkward, and I was also embarrassed by the way I looked. I was in my work clothes— stained sweatpants, a greasy T-shirt, and dirty sneakers. She passed me without recognition.

I was living a double life, and I liked it.

9

AFTER DAYS THAT FELT like an eternity, Chef Zhang finally allowed me to knead the dough. Since Chairman Wang had familiarized me with dumpling dough, I figured it would be a cinch.

Using a rice bowl, Zhang scooped the flour into a big tin container. He placed the container under the tap and turned on the water. (After my many lessons with Chairman Wang, I knew it would be futile to ask him the exact ratio of flour to water.) Then he set the tin bowl down and stood back to watch. As I immersed my hands in the mixture, I felt my confidence drain away. For one thing, the sheer quantity of dough was much greater than anything I had kneaded with Chairman Wang. And with a smaller proportion of water, the dough was tougher, which meant that I had to push with the entire force of my body, gritting my teeth.

The first task was to get everything to stick together. I had watched Zhang knead, with an artistry that made it look easy. Using a circular motion, he ran his hands lightly just underneath the surface of the flour. The dough began to clump, first in little florets, as random pieces adhered. He kept a bowl of warm water nearby so he could add moisture, trickle by trickle, as needed. When the dough started clinging into a single mound, he transferred it to a clean counter. This was where the muscle came in—he pushed down on the dough with his palms, standing on the balls of his feet so he could apply all of his weight to the task. When he was finished, he divided it into slabs, each as heavy as a newborn baby. He covered them with a damp cloth, since the dough would dry out if left exposed for more than a few minutes.

I divided my dough into smaller balls to make it easier to knead, but even so, I could not get it to stick together without it sticking to my hands. I'd finish with fingers coated in a gluey white substance, and for the rest of the day, in the lulls between customers, I'd pick at my fingers as if they were scabbed. Gradually, as my skills improved—I learned to sprinkle dry flour on my palms as I kneaded—my hands were almost as clean after kneading as they were before I began.

Even after I had mastered the kneading, however, Zhang refused to let me grate. I stewed over this for several days before I realized that it wasn't that he thought I couldn't do it. He thought it was too *buhao yisi*—embarrassing—to have a guest do the hardest job in the kitchen. He was always telling me to *xie'r yi hui'r*—rest for a while. "Why don't you relax in the dining area?" he would say. "I'll make you something to eat." One afternoon when the lunch shift was over, he filleted a whole carp into tissue-thin slices, the way I'd seen the chefs at the cooking school do it. He poured some soybean oil into the wok, and the residual water in the wok sizzled. As he waited for the water to evaporate, he gathered up a handful of chopped leeks, ginger, garlic, and pickled vegetables and held it over the fire. Standing motionless, he listened to the sounds of the wok.

Once the water had evaporated and the wok was silent, he released the minced ingredients, creating a cacophony of crackles and savory scents. When he threw in the fish, it seared with a hiss.

Zhang had mentioned that he was afraid that his banquet cooking skills were getting rusty with all the noodle grating, but his worry was unwarranted: he was still a maestro at the wok. The fish was tender, and the spicy-sour broth he stewed it in could have been a meal in itself, poured over rice.

I asked Zhang if he thought that running the noodle shop was harder than working at Yushan.

"Well, at Yushan we had twenty chefs, and we would serve two thousand customers during lunch," he said. "We were handling about a hundred customers apiece. There were a lot more dishes to worry about." He added, "I really should have left Yushan and come out on my own earlier. I would have gotten rich by now."

"It's not too late," I said.

"There are too many of us now," he said, echoing Chairman Wang's comments about migrants in Beijing. In the past, the government had restricted the number of migrants who came to the cities. But as cities boomed and increasingly depended on migrant workers (most urban dwellers weren't willing to take labor-intensive jobs), the government had largely given up on stemming the rural tide.

Haizi was bored by this serious talk. "Let's drink!" she said. She bragged that she had "learned" how to drink over the summer.

So Zhang brought out the *baijiu*, a Chinese grain alcohol, and a couple of bottles of beer, and we toasted our hard work.

FISH AND PICKLED VEGETABLE SOUP
(SUAN TANG YU)

¼ cup cornstarch
¼ cup all-purpose flour
1 pound carp fillets, about ten pieces
¼ teaspoon salt
½ teaspoon powdered chicken bouillon
¼ cup rice wine or sherry
¼ cup plus 2 tablespoons vegetable oil
½ cup Chinese pickled vegetables, minced
 (*suancai,* available at Asian markets)
1 tablespoon minced leek
1 tablespoon minced ginger
1 tablespoon minced garlic
1 quart water
¼ cup white vinegar
4 dried Sichuan peppercorns
4 dried red chili peppers, coarsely chopped

In a small bowl, combine the cornstarch and flour. In a separate bowl, combine the fish, salt, ¼ teaspoon bouillon, and 2 tablespoons of rice wine. Add the cornstarch and flour mixture to the fish and make sure the fillets are evenly coated.

Place a wok over high heat and coat the bottom with ¼ cup of oil. When the oil is hot, add the pickled vegetables, ½ tablespoon leek, ½ tablespoon ginger, and the garlic, and stir. Add the remaining rice wine and stir. Add the water and the remaining bouillon. When the soup comes to a boil, lay the fish fillets flat, one by one, in the wok. Continue cooking over high heat. After 3 or 4 minutes, add the white vinegar. Reduce the heat and simmer for 3 or 4 more minutes. Pour the soup into a serving bowl. Garnish with the peppercorns, chili peppers, and the rest of the ginger and leek. Add 2 tablespoons of oil to the wok and place it over high heat, then pour the hot oil over the soup, making sure to singe the garnish. Serve immediately.

◎

A month into my noodle internship, Zhang's wife and son came to Beijing for the first time since Zhang had moved to the capital. His daughter stayed behind to study for her high school exams. Having them around made the shop more pleasant. With his wife, Yao, taking the orders, things ran more smoothly. She had a wide jaw, big lips, and dark skin. Her soft voice and country accent made her Mandarin hard to understand. Meeting Yao made me wonder if Zhang had arrived in Beijing without the streetwise air he now possessed.

Nine-year-old Shiqiang was only a few inches shorter than his father, and more mature than his twenty-year-old cousin, Haizi. Zhang called him Erzi, "son." They had the same pumpkin-shaped head, expressive eyes, and protruding ears, although the son's cheeks were round and plump where the father's were chiseled. Zhang commanded Erzi to fetch boiling water for tea, carrying it back in tall thermoses. He delivered bowls of noodles to the workers in the furniture stores and bought his father cigarettes from a nearby tobacco shop. The rest of the time he sat in the mess hall reading his textbooks. He seemed to fear the kitchen.

"Do you know how to make noodles?" I asked him one day when he poked his head in.

"Nope," he said, and raced back into the mess hall.

"I won't let him. He's not going to be a chef. He's going to college!" Zhang declared.

Once in a while I would join Erzi at a cafeteria table and read passages in his literature book out loud with him. After years of studying Mandarin, I still couldn't read faster than a fifth-grader.

"In the beginning, heaven and earth were one," we read in unison. "The universe was a *hundun*."

I paused. "A wonton?"

"No, that word means 'very messy.' It only *sounds* like wonton," Erzi explained.

Clearly, I had spent too much time around noodles.

I knew I was getting closer to my chance at grating the noodles when Zhang showed me how to make the sauces for them. He started with my favorite, a savory pork sauce, which turned out to be amazingly simple. He infused a number of spices in hot oil. Then he added cubes of pork belly to the wok, allowing the meat to soak in the flavors from the oil. Soy sauce came next, the liquid sizzling and stoking the fire, which briefly leapt over the rim of the wok. Finally he added water and left the sauce to reduce for most of the morning. The eggplant sauce was just as easy. He used round eggplants that had a meaty taste, unlike the bland oval ones I had eaten in the United States.

CARAMELIZED PORK SAUCE (*ZHU ROU LU*)

1	pound pork belly, cut into 1-inch cubes
¼	cup vegetable oil
¼	cup sugar
2	tablespoons plus ¼ cup soy sauce
1	quart water
½	teaspoon salt
1	leek, white part only, cut into 1-inch pieces
2 or 3	thin slices ginger, thumb-sized
4	star anise
2	bay leaves
1	tablespoon powdered chicken bouillon
2	whole dried red chili peppers

Fill a wok halfway with water and bring to a boil. Add the pork and blanch it for 2 minutes. Drain and set aside.

Combine the oil and sugar in the wok and place over high

heat. Stir vigorously until the mixture begins to caramelize and turn brown. Add the pork, stir, and add 2 tablespoons soy sauce; stir vigorously for 2 to 3 minutes. Add 1 quart water, ¼ cup soy sauce, salt, leek, ginger, star anise, bay leaves, bouillon, and chili peppers. Simmer for 30 minutes to 1 hour and serve over cooked Chinese noodles.

EGGPLANT SAUCE (*QIEZI LU*)

¼ cup vegetable oil
1 tablespoon minced leek
1 teaspoon minced ginger
2 teaspoons minced garlic
1 large eggplant, cut into ½-inch cubes
1 green chili pepper
1 white potato, cut into ½-inch cubes
1 tomato, diced
¼ cup soy sauce
1 cup water
½ teaspoon powdered chicken bouillon
⅛ teaspoon salt

Add the oil to a wok and place over high heat. When the oil is hot, add the leek, ginger, and garlic, stirring for 1 minute to coat them with oil. Add the eggplant and stir for 2 minutes so the oil coats each cube, adding oil if necessary. Add the pepper, potato, and tomato and stir for 1 minute. Add the soy sauce and water, reduce the heat, and simmer for 2 to 3 minutes. Add the bouillon and salt and simmer for another 2 minutes. Serve immediately.

Sure enough, a few days later, after the lunch crowd had subsided, Zhang asked if I wanted to give the noodle-grating a try. He handed me the dough and the boxy, handleless knife and uttered a mantra that he later repeated whenever I grated:

"The knife doesn't leave the dough; the dough doesn't leave the knife." As instructed, I cupped the board that the dough sat on just below my shoulder, as if it were a violin. When Zhang stepped outside to smoke a cigarette, Auntie Xu crept into the kitchen to watch. "You're doing it wrong! You can't cut them right!" she howled, lunging for the dough. I doggedly held on. Thankfully, Auntie Xu gave up; had she been persistent, she would've had to wrestle me to the floor for control of the dough. The grating reminded me of the way it had felt the first few times I chopped with a cleaver; I didn't have the rhythm down. After a few dozen syncopated strokes, punctuated by splashes of the scalding water, I finally found an angle at which the ribbons of dough plopped seamlessly into the water.

I put the slab down. My left arm quivered from the weight of the dough; my right arm had stiffened with repetitive motion. It had taken an excruciating three minutes to grate one bowl's worth of noodles. I picked up the slab and tried again. After five three-minute intervals that reminded me of the physical fitness tests in high school gym, I had produced two dollars' worth of noodles, enough to fill five bowls.

"Not bad," Zhang said when he returned.

Auntie Xu forced a laugh. "She's not bad, is she?"

I wasn't terrible, but I wasn't all that great either. I was diligent, though. The repetitive grating motion was similar to a tennis backhand, minus the follow-through. I was a decent tennis player, but the mental part of the game killed me. In the kitchen, it was the exact opposite. I had the sheer will; I just didn't have any natural noodle talent.

I suffered through eating what I had grated. The noodles were supposed to be chewy, even strands that your teeth bit into and released in a steady rhythm. My noodles lacked a consistent texture. Fat strands of raw and gummy dough were interspersed with stringy pieces that tasted limp and overcooked. Eating a bowl of my noodles was like tucking into a half-burnt, half-undercooked pizza.

In fact, noodles could burn if boiled too long, as I discovered one day when I saw Zhang bent over with his nose in the wok. He scooped up a few strands with a slotted spoon. "Smell," he said. I leaned forward and inhaled a charcoal scent.

Determined to achieve greater mastery, I grated away, day after day, ignoring Zhang's warnings to take it easy. After a couple of weeks, I understood why I should have heeded his advice. My right elbow and forearm were constantly sore, the pain pulsing down to my middle and ring fingers. It hurt to bend my wrist. A friend dubbed the condition "noodle elbow."

Aside from the traditional knife-grated variety of noodles, Zhang offered two other kinds: cat's ears (*mao erduo*) and hand-rolled noodles (*shougan mian*). He offered to teach me how to make them, and while I suspected that he was only trying to give me a reprieve from the knife-grated noodles, I was glad to learn.

The same dough was used for all the noodles. For the cat's ears, he rolled out the dough, stopping when it was twice as thick as a pie crust, and cut it into pieces the size of eraser heads. He pressed down on the dough with his thumb, rolling it from one side to the other. The dough magically morphed into oval shells that resembled orecchiette pasta.

The trick with the hand-rolled noodles was to roll the dough into a long, super-thin sheet. Zhang worked it around the rolling pin, sprinkling it liberally with flour as he pressed and rolled it on the counter to prevent it from sticking to itself. When the dough was thinner and softer than newsprint, he released it, spinning and pivoting the pin so that the dough folded over itself into many layers. With a knife, he cut the folded dough into long, thin noodles.

Both the cat's ears and the hand-pulled noodles were less labor-intensive than the knife-grated variety, but Zhang priced the knife-grated noodles 12 cents cheaper than the variations. Knife-grated noodles represented his roots, and I suspected he took pride in sweating over the wok with a hunk of dough in his hand. It was a perverse pleasure that I was developing, too.

ZHANG'S NOODLES

Place 4 cups of all-purpose flour in a large bowl. Stir in 1 cup of water, and mix the water and flour with your hands. Slowly add more water, about ¼ cup at a time, mixing thoroughly until the dough is springy and firm but not dry (it should be firmer than dumpling dough). Transfer the dough to a clean counter and knead for 3 to 5 minutes. Cover with a wet cloth and let it sit for at least 10 minutes.

KNIFE-GRATED NOODLES (*DAOXIAO MIAN*)

You'll need a special knife with a curved blade to make these noodles. (This knife—called a *daoxiao mian dao*—will be hard to find; if you live near a Chinatown, try a kitchen supply store there.) Knead Zhang's noodle dough vigorously, shaping it into a loaf approximately 8 inches long, 4 inches wide, and 3 inches thick, making sure the edges are rounded. Bring a pot of water to a boil. Stand over the pot and hold the knife in your dominant hand and the noodle loaf vertically in your other hand, supporting it with your palm. Starting with the knife's edge at a 45-degree angle to the dough, slide it along the length of the dough from top to bottom, grating one long piece at a time, about ½ inch wide. Start on one side of the dough (on the right if you're right-handed; on the left if you're left-handed), and once you have completed one vertical downward motion, move your knife ever so slightly left or right and continue grating as if you're peeling a carrot. Grate one serving (about 50 strands), boil the noodles for 3 minutes, drain, and serve immediately with sauce.

CAT'S EAR NOODLES (*MAO ERDUO*)

With a rolling pin, flatten Zhang's noodle dough into a sheet about ¼ inch thick. Cut the sheet into ½-inch-wide strands, then cut each strand into ½-inch-long pieces. Sprinkle flour over the dough pieces. Place your thumb on a piece of dough and roll from one side to another, so it curves into the shape of a cat's ear, or shell. Repeat for all the pieces. Boil the noodles for 3 minutes, drain, and serve immediately with sauce.

HAND-ROLLED NOODLES (*SHOUGAN MIAN*)

On a large floured counter, flatten Zhang's dough with a rolling pin, making a very long, thin sheet, about ¼ inch thick and no wider than the width of the pin. (To prevent sticking, lightly sprinkle the dough with flour.) Then, starting with the end that's closest to you, flatten the dough to the thickness of newsprint, sprinkling liberally with flour, and begin to wrap the paper-thin dough around the pin. As you wrap, sprinkle flour on the dough so the layers don't stick together. When all the dough is wrapped around the pin, slowly unroll it so that it folds over itself in layers about 3 inches wide. Cut the dough crosswise to form ⅛-inch-wide strands. Boil the noodles for 3 minutes, drain, and serve immediately with sauce.

◎

Whatever my progress with the noodles, I was fast mastering the attitude of a typical Chinese chef. One day during lunch rush, a customer brought her noodles back to the stall.

"There's a hair in this!" she said, thrusting out her bowl.

Zhang rushed to the counter. The strand of hair was long and thick, not unlike mine. Zhang apologized and said a new bowl would be quickly on its way.

I felt more indignant than embarrassed. I had seen plenty of hair in my food in my years of living in China. The first few times, I had been horrified. Then I was merely annoyed. Eventually, I simply picked them out or ate around them.

"The nerve of that woman!" I said to Zhang when he returned to the kitchen. Here we were, sweating away, while she was paying only 40 cents for her lunch. And it was, after all, just my hair.

Next door to Zhang's stall, three Sichuanese siblings ran an efficient stir-fry stand that did a brisk business. The rest of the canteen's booths were vacant. A sign that read SEAFOOD STIR-FRY hung above one of the empty stalls. Before Zhang arrived, a family from the coastal province of Fujian had operated that booth, but it closed abruptly after the father had gambled away all their earnings. The family left without paying the rent, the gas bill, the chef they had employed, and their food suppliers.

"Even the newspaper subscription was left unpaid," said Auntie Feng as the lunch shift began one morning.

"Really?" said a customer as she waited for her noodles. "We used to eat there every day."

"Was it any good?" I asked the customer.

"How do I put it? It was a fast meal. It's all about the same." She shrugged; it was a canteen. Did anyone expect gourmet food in a canteen? No matter how good I thought Zhang's noodles were, they were going to be underappreciated in such a setting.

A few weeks later, Auntie Feng quit to take a job at a drug factory. A mute man in his fifties replaced her. With his rumpled hair and green army fatigues, he looked utterly lost, as if he had just been fighting in the Communist Revolution and somehow managed to time-travel into the present. He requested bowls of hot soup from Zhang and the Sichuan family and sat with a newspaper in the Fujian family's empty booth. But instead of reading, he tapped his hands against the counter and stared into space.

The Han brothers couldn't wait any longer to find new tenants and decided to set up a stall of their own. They hired two chefs, put up a huge red menu on the wall, and worked the counter themselves.

"Is this going to affect your business?" I asked Zhang.

He shrugged. "Not really. Their menu is more similar to the Sichuanese family's. People already know what they're going to eat when they come in. If they want noodles, they'll come here. If they want rice and stir-fries, they'll go over there."

The Sichuanese family's anger was palpable. The woman who ran the stand stopped saying hello to me when I arrived in the morning. As the lines shortened at her stall and grew at the Han brothers' stand, her scowl deepened. Her bright yellow apron clashed with her dark, beady eyes that narrowed at the booth next door.

"That woman has a *suzhi* problem," Zhang said. The insult was my favorite Chinese expression. *Suzhi* meant "quality," and when it was used to describe someone, I envisioned a defective person rolling off a Chinese assembly line.

In this case, though, it seemed to me that both sides had *suzhi* problems. The Hans had copied the Sichuanese's menu, after all. Soon, the Sichuanese began stealing bowls and plates from the Hans. The brothers—who, the Sichuanese seemed to forget, were also their landlords—then demanded their rent money before it was due. The Sichuanese, annoyed that Zhang had chosen to stay out of the battle, decided to launch an all-out assault and began making noodles that mimicked Zhang's.

"They can make whatever they want," Zhang said coolly. "I'm not going to fight it."

One afternoon, a shouting match erupted between the Sichuanese and the Hans. The brothers told the Sichuanese to pack their things and leave. The next day when I came to the canteen, the booth was empty. The younger Han brother was sorting through what little the family had left behind.

"Their lease was up," he said as he removed the rice cooker from the bare kitchen.

Finally, after almost three months of working at Zhang's shop two or three days a week, I mastered the noodle.

Zhang had let me take over the second half of the lunch shift. He lit a cigarette and watched me grate.

"You've got the hang of it," he said with a hint of surprise.

And I had. The movements came fast and easily. I was on autopilot. The noodle knife didn't get stuck halfway through the slab. The noodles no longer looked as if they had been hacked by a machete; they looked like proper ribbons. I grated a bowl, then rested for a moment, then began another bowl. Within three minutes, I had made three bowls. I was never going to be as fast as Zhang, but I could grate well enough to handle a slow stream of customers.

My forehead was covered with sweat as the lunch shift wrapped up. I was too tired to rejoice much, but I felt a satisfaction I had never experienced before. Things I had made with my own hands were being bought and eaten. I was a noodle maker.

10

ONCE I HAD MASTERED the noodle shop's entire menu, it was time to move on.

I thought about trying to find an "authentic" Beijing restaurant, but the city had never really developed a full-fledged cuisine of its own, and could claim only a few regional specialties, like Peking duck, *zhajiang* noodles, and the suspiciously foreign-sounding Mongolian hot pot. With a constant influx of people from other regions and a climate that wasn't conducive to growing much, Beijing—a hub like New York or London— consumed the best of what originated elsewhere. But dumplings were a genuine Beijing specialty, even if few people (Chairman Wang aside) bothered to make them at home anymore. Beijingers still craved them, which was why Xian'r Lao Man was a success.

The dumpling house was a five-minute bicycle ride from my home. During the lunch and dinner hours, the place bustled like a crowded New York diner: elderly couples sat next to hipsters in thick-rimmed glasses and T-shirts with abstract designs. People crowded around the door, waiting for tables to open up. The owners had decorated Xian'r Lao Man with replicas of Chinese antiques and black-and-white photographs, which made the atmosphere more pleasant than many of its humble neighboring restaurants.

The dumplings, like the setting, had been given an update. They came with a choice of sixty different fillings. My favorites were cabbage and peanut (the peanuts gave them a nice crunch), pork and corn (I liked corn in anything), and shrimp and Japanese tofu (a silken variety with a smooth, pudding-like texture). The restaurant also offered Chinese salads, which were nothing like the "Chinese" chicken salads I had seen in America. Carrot, cabbage, bean sprouts, and shredded tofu were tossed with vinegar, sugar, and a dash of sesame oil.

DONGBEI SALAD (*DONGBEI DAPAN CAI*)

½   medium head of green cabbage, finely shredded (about ½ pound)

1   cucumber, cut in quarters lengthwise, seeded, and thinly sliced

1   carrot, shredded

¼   cup roasted peanuts, shelled and skinned

1   large handful of bean sprouts

¼   pound shredded tofu (*doufu si* or *gansi*, available at Chinese groceries)

2   teaspoons Chinese black vinegar

2   teaspoons leek-infused vegetable oil (see note)

¼   teaspoon salt

1   teaspoon sugar

In a large bowl, toss all the ingredients and serve immediately.

Note: To make leek-infused oil, heat the oil in a wok until it steams. Cut the white part of a leek into 1-inch slices and add to oil. Just before the leek begins to burn, remove the wok from the heat and strain the oil. It is now ready to use.

The owner was a friend of a friend. Boss Sun had a pixie hair-cut and a delicate, narrow face. She powdered her face white, and she was fond of outlining her lips with a sharp red liner. Though she was forty-two years old, she dressed as if she were two decades younger. One afternoon, she sauntered into the kitchen in a T-shirt and a pair of tight jean shorts, a cigarette hanging from her mouth as she counted a thick roll of hun-dred-*yuan* notes. Another day, she romped around the dining room in a blue babydoll dress and leggings. I wasn't surprised when she told me she had started her career as a beautician.

Along the way, though, she switched fields and became a real estate agent. One day she was browsing through property listings when she found a space that seemed perfect for a restaurant. Like many Chinese, she and her husband, a graphic designer, liked to conduct business in restaurants, and they had daydreamed of opening a place that would serve dumplings like the ones from their childhood. She put her husband in charge of designing it. It had opened three years ago and im-mediately attracted a local following, despite the fierce compe-tition in a city filled with restaurants.

"The reason we're successful is that we've kept our prices low," said Boss Sun. No item on the menu cost more than 40 *yuan*, or $5. "Our competitors can't charge so little because they have to pay rent. But we own our building."

Boss Sun and her husband had two grown children, seven-teen and twenty-one years old. Two years earlier, she'd gotten pregnant again and had brashly decided to violate the one-child policy once more. Technically, there wasn't a fine for having ad-ditional children; the government simply refused to give them identity cards, which they needed for school and employment, unless their parents paid a large fee, ranging from a few hundred to a thousand dollars. Like many parents, Boss Sun was postponing the payment of the fee until her daughter reached school age. Sometimes she brought the identityless girl to the restaurant and played with her in a corner. "Since the third baby, I just don't have as much energy as I used to," Boss Sun said one night. I had joined her for a late dinner, and a number of times our meal was interrupted when her cell phone rang or she abruptly bounced off to the kitchen. She moved about with the energy of someone who had consumed several pots of strong tea. I wondered what she had been like before baby number three.

During a lull when she sat back down, we gossiped about our mutual friends, a foreign couple who lived in the neighborhood and came to Xian'r Lao Man a few times a week.

"You can tell they get along well," Boss Sun said, sighing. "Foreign men are much more romantic than Chinese men. Sure, my husband and I talk all the time, but all we talk about is business. Chinese men don't express themselves. They boss us women around."

Not all foreign men were romantic, I said. American wives, I assured her, complained about their husbands plenty.

Did I have a *laogong*, a husband? she asked.

No, I said, but I had recently started dating someone.

"Is he a foreigner?" she asked.

I told her yes, he was an American.

Her eyes narrowed. "Is he romantic?"

Craig, the man I was dating, happened to be a hopeless romantic, I conceded. She smiled smugly. That cemented it for her: all foreigners were romantic. She said excitedly, "You really should have kids soon! The longer you wait, the harder it will be."

"It's a big responsibility," I said. Besides, I added, we hadn't been dating that long.

"Why not just do it? Wouldn't that make your *laogong* happy?" she said. Many women used the word to describe their boyfriends, which I found unsettling. "You can give your baby to your mom to raise." Boss Sun's mother often took care of her third child, the way lots of Chinese grandparents did. Since the parents were busy with their careers and the grandparents didn't have much to do, why not?

Boss Sun didn't spend that much time in the restaurant, though. She blew through the kitchen occasionally, sometimes pausing to wrap a few dumplings, just for kicks. She left the management of the dumpling room to Sister Cao, a short woman with arms as thick as a weightlifter's from years of dumpling-making.

The heavy woks in the main kitchen were commandeered by men, but the dumpling room was an all-female operation, with workers ranging from twenty to forty. They stood elbow to elbow, rolling skins and wrapping the orders that were shouted every few seconds at the height of the lunch and dinner rush. On an average day, the dumpling room went through 55 pounds of ground pork and 250 pounds of flour. In the morning, a chef would fill a large tin bowl with pork. She'd put her hand in the bowl, which was as big as a truck tire, and move her hand like the beater of an electric mixer until the pork had the texture and color of goose liver pâté. This pork wasn't the main filling in the dumplings; it was the glue that held a mostly vegetable filling in place. I remembered how Chairman Wang was horrified at how much meat I put in my dumplings. I had violated a custom that spoke to her past: meat wasn't traditionally a main dish in China; its purpose was to enhance the flavor and texture of vegetables, which had a more central role. Though meat consumption had risen considerably, somehow dumplings were left unchanged, like a relic.

I thought of myself as a pro with dough by this point, after wrapping dumplings with Chairman Wang and grating noodles with Chef Zhang. But working with 250 pounds of flour every day was a different story. The flour was an expensive brand called Snow Flower. After it was mixed by hand with water, it was placed in a machine for additional kneading, then set aside for half an hour to let it *xing*, wake. Then the dough was worked into slabs, and the slabs were cut into smaller sections that were rolled into snake-like ropes, the way I had learned from Chairman Wang. But rather than cutting the dough further, the chefs snapped off little pieces, making balls —bigger than a gumball, smaller than a ping-pong ball. During the lunch shift, the chefs worked through 15 slabs of dough. Each slab was cut into 150 balls. That made a total of 2,250 dumpling wrappers per batch of dough.

"Why don't you use a knife?" I asked one chef.

"Once you're good at it, your fingers are faster and more consistent than a knife," she said. "Your thumb will tell you how much dough you need to cut."

I tried to listen to my thumb, but the balls of dough I snapped varied in size. I moved on to rolling out dumpling skins. The chef next to me rolled them out two at a time, using her hands to flick the skins ever so slightly each time she pushed the rolling pin over them with her palms. When the skins had made a full revolution, she flung them like Frisbees onto a mounting pile of wrappers across the counter. In the time it took me to roll out a single skin, she had made four.

Since I couldn't compete on speed, I concentrated on making my skins rounder than my neighbor's. Mine were rounder than everyone else's, I thought with satisfaction.

Then the woman on the other side of me shrieked, "She's doing it wrong!"

The whole dumpling line stopped and looked up. I almost expected an alarm to go off.

What could possibly be wrong with my perfect circles? I wondered.

My neighbor held one of hers in her palm, bringing it up to eye level. "See?" she said, pointing to the center. The center was slightly thicker than the edges. Up close, it resembled a flying saucer.

As the lunch rush came to a close, the dumpling makers turned to preparing the fillings for the evening. Chefs gripped knives in both hands, mincing giant bunches of celery at a galloping tempo. Then the heady smell of celery was replaced by the scent of licorice that came sweating out of the fennel. The restaurant made its own sausage, a chef using a carrot to shove the ground meat scraps into the casing of pig's intestine through a funnel made from the sawed-off top of a plastic soda bottle.

As the chefs engaged in these repetitive, labor-intensive tasks, I gazed around, thinking, Where's the Cuisinart? Didn't we invent a machine for that? Everywhere, arms and hands were swinging and stirring, moving like levers and motors, pushing the food down the human assembly line. I was in a dumpling factory.

In the factory, there wasn't time to do the careful pinches I had learned from Teacher Wang. The mass-produced dumpling had to be stuffed, pressed, and sealed in a matter of seconds.

Sister Cao kept a watchful eye over me when I tried my hand at wrapping. She didn't like having a visitor in her kitchen, especially a clumsy one. She scowled as my dumplings ruptured and spilled their fillings, like wounded soldiers on a battlefield.

One afternoon, unable to take it any longer, she came over and pushed me aside. "You don't wrap them the same as us," she said frostily, picking up where I had left off.

She placed a dumpling skin on her palm. Using a wooden stick, she deposited a gob of filling on the skin. She folded the dumpling into a half circle, her thumbs and index fingers clamping the edge to seal it in one swift motion. When she released it, the dumpling resembled a ping-pong ball with wings.

"Everything has a method to it, doesn't it?" she said in a no-nonsense tone. My reject dumplings remained on a tray on the counter, mocking me.

I had an idea: rather than practicing on dumplings that would go out and horrify the general public, I decided to buy dozens of dumplings and wrap them myself. Even Sister Cao couldn't object to that.

The wrapping took on a meditative quality, and gradually my dumpling-making improved. I was relaxed enough to eavesdrop on a conversation between two of the wrappers. They looked to be in their thirties, and both had left their small towns to come to Beijing.

"Your land is too small," said the one from the nearby province of Hebei. She wore heavy eye makeup that made her look like a raccoon. "Where I'm from, we get seven hectares of land per household."

"We get three hectares," said the pudgy one with the heart-shaped face. She was from a western desert province called Gansu, China's equivalent of, say, New Mexico.

"Our yard is very big," said the Hebei woman.

"Ours is all concrete," the Gansu lady replied.

"We grow grapes, persimmons, cotton, and flowers."

"We don't grow any flowers. Our land is too small."

"The places we are from are very different."

"They recently built a highway and took away more of our land."

"When there is no land, you have to leave and go work somewhere else."

"That's why people go into the towns to do business."

"But your brain can't rest when you are selling things."

"I like to do business. I used to sell hats and shoes."

Dumpling wrappers were generally on the low end of the restaurant totem pole, a notch above the dishwashers and a notch below the waitresses. Boss Sun gave them the usual couple of days off each month and paid them $100 a month, plus free room and board. She seemed to think that was generous. "Why not give them a bit more and make them happy?" she said. But it was still lousy pay, and she found other ways to scrimp. Their lunch was typically rice, tofu, and cabbage that was on the verge of rotting. Rarely did they eat meat. Boss Sun didn't let them eat the dumplings they made—she considered the humble dish too luxurious for her staff. The workers came from everywhere but Beijing, from the northern coastal areas to the inland deserts. Most Beijingers refused to take a job as lowly as dumpling-wrapping, which they saw as a job for migrants. Bosses didn't hire diversely for diversity's sake. As a Beijing boss, she didn't want employees with similar allegiances, because that made it easier for them to come together and complain if they weren't happy.

One afternoon, Boss Sun stormed into the dumpling room.

The dumpling orders were taking way too long—one table had been waiting for an hour. Dumplings were going out with broken skins. One order was short five dumplings.

"Who's the one from Gansu?" Boss Sun screamed.

The portly Gansu lady, who was washing a tray in the sink, raised her hand.

Boss Sun began scolding her in a harsh tone.

"Why are you blaming me?" the Gansu lady asked.

"It must be your fault because you're from Gansu, and people from Gansu can't understand Mandarin properly," Boss Sun said.

Sister Cao cleared her throat. "Actually, she's not the problem. It's those two over there," she said, pointing across the room.

Though Sister Cao had fingered two people nowhere near me, I had the nagging suspicion that I was to blame. After all, the fillings still spurted out of every other dumpling I folded.

It was hard to get to know any of the women on the dumpling line, since things were always moving so fast. But Hu Guirong, the Hebei lady with the raccoon makeup, stuck out. She often gazed into space with a dreamy look on her face as she rolled out dumpling skins. Occasionally I looked up to see her staring at me. When our eyes met, she smiled shyly and looked away.

Hu had long, frizzy hair and thin lips. On closer inspection, I noticed that it took numerous coats of mascara and dark eye shadow to achieve her mask-like appearance. I guessed she had been pretty in her youth, but the years had given her a weather-beaten look. While we rolled and wrapped one day, I asked her the usual questions that strangers asked of each other, unsure how else to break the ice. Was she married? Did she have kids?

She looked away, then said awkwardly, "I have married. I have a child." Then she put her head down and paused. "I'll tell you about it later."

Another afternoon, she said quietly, "I hear you are from America. Do you believe in Christianity?"

I told her I wasn't religious, but that some of my relatives were.

After we had wrapped several orders, she whispered, "I'm Christian. But the boss doesn't give Sundays off, so I can't go to church."

Later, she asked if I had ever been to Taiwan. I told her that I had many Taiwanese relatives, and that I tried to visit once a year to see my grandparents in Taipei. "Really?" she said. "I've always wanted to go there."

Our conversations were no more than snippets. One of us would say something. Then a rapid succession of orders would come in. During the next lull, we'd exchange another few words. Two questions and answers, three batches of dumplings, one afterthought, a new topic sentence.

After a few days of stunted conversation, Hu followed me out of the restaurant at the end of a lunch shift.

"I wanted to explain why I didn't tell you more about my family," she said. "The truth is, I divorced my husband. He was a police officer, and he was never at home. He was cheating on me, and I couldn't take it anymore. When I divorced him, I had to move out, and he got custody of our son. So I left. Nobody in the kitchen knows."

Divorce was becoming more common in China, but in a society where the family was more important than the individual, it still carried a stigma, especially for women. In Chinese divorces, often it was the husbands who got custody of the children.

"I have several friends who are divorced," I said. "It's okay."

As we approached an intersection, there was an uncomfortable pause.

"That's all I wanted to say for now," Hu said. We said goodbye, and I turned at the corner to head back to my apartment. Hu walked back to the restaurant, where she'd soon start the dinner shift.

I didn't work at Xian'r Lao Man for long. The repetitiveness of the job depressed me. Hu didn't stay there much longer either.

"I'm never working in a restaurant again," she told me just after she quit. "It's bitter work. Two days of rest a month isn't enough. Everyone's replaceable. New people come and go all the time. It's too capitalist."

My contact with Hu was erratic after that. She didn't have a cell phone, and she was constantly traveling. She said she had to go down south to take care of some business, without elaborating further. I had given her my phone number, and once in a while she would call and ask, "Do you still remember me?"

Months later, on Christmas Eve, Hu and I met up at Chongwenmen Church in downtown Beijing. A line a quarter mile long stretched out from the building, as Chinese Christians waited for churchgoers from the previous service to file out. Hu had already been to an earlier service, and neither of us felt like standing in the freezing cold. We sought shelter in a small shack that served snacks, and I ordered two *roujiamo,* a specialty of northern China that was like a sloppy joe: steaming hot buns filled with chopped pork, spicy green peppers, and coriander.

I marveled at how many people had been waiting to go into the church. "They don't take their religion for granted here," Hu said. She was dressed in a puffy green jacket and a black baseball cap. She was wearing mascara, but no eye shadow. The less makeup she had on, the prettier she looked.

As we sat down to eat, Hu told me her story. She had become a Christian in 2001, after she left her hometown and went to the southern coastal province of Fujian. There she met a Christian woman who became something like an older sister to her.

"I was Buddhist before. I was always spiritual. But I liked Christianity because it seemed more modern than Buddhism," she said. This was characteristically Chinese thinking, associating Western ideas with progress and Chinese beliefs with tradition.

In Fujian, Hu was exposed not only to Christianity but to the possibilities of the outside world. The region was a notorious hub for human smuggling, and she met "snakeheads" who promised they could arrange a fake marriage with a Taiwanese man, ensuring her a visa to the island.

"I had heard that Taiwan was very modern and developed and that you could make six thousand Hong Kong dollars a month," she said. "I wanted to go somewhere freer than China."

Hu was also looking to escape her past. Increasingly, her ex-husband cut short the amount of time she could speak to her son by telephone. After he remarried, his new wife forbid the son to speak to Hu at all. In 2003, with the money she had saved selling fruit in Fujian, she paid a snakehead $6,500 to help arrange a marriage with a fifty-year-old man from the Taiwanese countryside. "He was very *tu*," she said, using a word that meant "uncouth." "We spoke on the phone, and I could barely understand his dialect. But I didn't mind."

After she paid the money, the snakehead said the deal had fallen through. Her money was never returned. I resisted commenting on several details in the setup that struck me as fishy. I didn't say that Taiwan had its own currency, so it was strange that the snakehead would talk about salaries in Hong Kong dollars. I didn't tell her that everyone I had ever met in Taiwan spoke comprehensible Mandarin. I didn't point out that it was foolhardy to pay the money in advance. I didn't think Hu would appreciate my belated advice.

She got scammed a second time, and a third. The second man was supposedly a cripple who needed a wife to take care of him. The third man was obese; she had met the supposed match at a hotel before the deal fell through. In total, she had spent $25,000.

"That's why I went down south after quitting Xian'r Lao Man," she said. "I wanted to see if I could get any of my money back. But they said it would be impossible." There was no legal recourse, and Hu said she had finally given up hope of going overseas.

After the somewhat oppressive conditions at Xian'r Lao Man, I looked forward to visiting Zhang back at his small operation toward the end of the summer. Though his work wasn't so different from that of the dumpling factory, it was at least done on a human scale, and he was his own boss.

I discovered that Haizi was no longer there. As the summer drew on, Zhang told me, things had soured between him and his niece. He no longer had as much patience for her ineptitude, and often by the middle of the lunch shift she looked as if she was about to dissolve into tears in the already steamy kitchen. Zhang's wife and son were leaving soon too, so that Yao could attend to their daughter, who would be entering high school back in their hometown. Besides, Yao didn't like Beijing. It was too polluted, and the room they lived in was too muggy to sleep comfortably. "The walls are hot, even at night," she said.

"Won't you miss your husband?" I asked.

Yao giggled like a schoolgirl who had been asked about a secret crush. Then she shrugged. "We're used to it. We got married a long time ago. I met him when I was eighteen, got married when I was twenty-one, and had my first child when I was twenty-two." She had encouraged Zhang to go to the capital to build his career and make money he could send back home. There weren't enough opportunities "to develop oneself" in the countryside.

Her unromantic attitude didn't surprise me. Aside from a few short trips Zhang had made home to Shanxi, the couple had been physically separated for eight years. Zhang sometimes spoke more to me in the course of a day in the kitchen than he did to his wife. She didn't seem to mind. I had heard his take on matrimony. "There's nothing good about marriage. It's too much trouble," he'd said. "Don't do it." But I suspected that, underneath it all, Zhang and Yao shared private moments of affection, something that was confirmed when one evening, after Yao had returned to Shanxi, I overheard Zhang talking to her on his cell phone in hushed, tender tones.

Without his wife in Beijing, Zhang wondered if he should continue plugging away at the canteen. He was only breaking even, and his rent, which he paid in advance, in three-month installments, was due in a few days. He had gotten a couple of offers to work in restaurants. If he gave up the stall and took a steady job, he could use the money that would have gone to his rent to pay for his daughter's schooling.

There was another opportunity brewing. Old Wang, who had worked with Zhang at Yushan, wanted to start a restaurant with him. Old Wang was in his forties, and had a classic Beijing face—wide forehead and big eyes. His hair was shaved in a buzz cut, and he had the build of a linebacker. Over two decades, he had worked his way up to senior management at Yushan, but had been laid off when new superiors came in. Unlike Zhang and Qin, though, he had worked long enough to collect a pension.

One morning, I found Zhang leaning against the counter combing through *Hands Lifting Hands,* a classified-ads newspaper, while Old Wang sat at a table near the stall.

"How big do you think this space is?" Old Wang said, glancing around the canteen.

Zhang estimated that it was around three thousand square feet. They didn't need anything that big, they agreed. Zhang pointed to a tiny ad in the paper. They both squinted at the patch. "This one says six hundred square feet."

Zhang went back into the kitchen to check on the stove. "He doesn't understand business," Old Wang told me. "He doesn't have a long-term vision. He just sees that if his costs are this much today, he better make that much, or maybe a little more." He sighed. "But his noodles are really good. There's a demand for this type of thing in Beijing."

Zhang reemerged from the kitchen and copied down some phone numbers from the paper. He gave them to Old Wang, who put on his sunglasses and left.

"Old Wang is very clever," Zhang said with admiration as the Beijinger bicycled away. "Yushan didn't like that. That's why he got laid off. But it didn't matter. He didn't want to work for the Communist Party anymore."

"Do you trust him?" I asked.

"It doesn't matter. I'm a migrant. As long as he trusts me, we're okay."

Zhang added that Old Wang had lent him $1,200 to help him open his first stall, the one where I had first tried his noodles. Though it was customary to borrow money from friends and relatives in China (banks rarely granted small-business loans), Zhang was surprised that Old Wang helped him. "I asked him, 'You're not afraid that I'm going to run away with your money?' He told me that he knew I was trustworthy."

Zhang's view of himself as a second-class citizen made me think of my taxi ride to the shop that morning, which had taken me along a dilapidated old street called Left Peace Road. Trash filled the gutters. Storefronts advertised bicycles, mobile phones, and sexual services, which could be bought at any hour from women in tight-fitting clothes who sat in a room lit with soft pink light and marked with a spinning barbershop pole.

"Look at this place," the driver had muttered. "Full of migrants. Look at how disorderly it is."

"Do you dislike migrants?" I asked him. Beijingers often blamed migrants for petty crime and other social ills, in much the same way that Americans held recent immigrants, legal and illegal, responsible for a host of problems.

"There's nothing not to like," he said. "It's just that they have lower *suzhi*." That word again. Just as he spoke, a bicycle swerved in front of his car. He stomped on the brake and honked, longer and louder than necessary. "Look at that. A migrant," he sneered, shaking his head.

. . .

As Zhang was making his third attempt at operating a restaurant in a year, nineteen-year-old Qin was just trying to stay afloat. She was on her own now, apart from the friends she'd trained with at Yushan. Through some Sichuanese contacts, she'd landed another waitressing job, across town at a Cantonese restaurant. Her pay was $100 per month.

I brought her a Chinese edition of *Elle* magazine. "*Ma-dang-na*," she said, reading the Chinese name of the woman on the cover. She had never heard of Madonna. Britney Spears didn't ring a bell either. She liked Chinese pop stars and rattled off a series of names that sounded as alien to me as Madonna and Britney Spears sounded to her.

"Let's go eat!" she said. She led me to a small home-style Hunanese restaurant and ordered *maoxue wang*, which loosely translates as "cat's blood bounty." It wasn't cat meat or blood, but it was exotic enough: the innards and blood of pig and cow with wood ear mushrooms floated in a spicy red gravy. The offal gave the dish a bold, hearty flavor, and we poured it over rice to tone it down a bit. The intestines tasted slightly rancid, but I liked the dish more than I expected.

I asked Qin if the adjustment had been hard.

"I knew something was going to work out," she said brightly. "I'm very adaptable. I've slept well since the first night."

But as we continued to eat, Qin seemed to grow more uncertain. Her plan was to save enough money to pay for training to become a tour guide, but that looked like less than a sure thing. She didn't know if she wanted to stay at the Cantonese restaurant, but she didn't have much of a choice. Her relationship with Yushan was too strained to allow her to return, and she had ruled out going back to Sichuan because salaries there were lower than in Beijing. She was thinking of going to Shenzhen, she said. Her mother, who had left Sichuan when Qin was eleven, was working in the southern boomtown.

When we finished eating, Qin slipped money to the waitress before I thought to reach for my wallet. Embarrassed, I chased her around the restaurant, trying to stuff money into her pocket. Qin giggled and ran away, as if we were playing a game of tag.

"You can pay next time," she said.

We had some time before Qin had to report back to the restaurant for the dinner shift, so we stopped by the guesthouse where she was renting a bed for a dollar a night. Her room was in the basement, down half a flight of stairs that led to a dim, narrow hallway. The basement was dark and dank, and the doors to the rooms were open. To the left I saw a woman holding a baby in her arms, and to the right a group of men loitered in a smoky, windowless room. Qin's room had four sets of bunk beds, on which three women, whom Qin had just met, napped while a small television droned.

Still, even in this dismal setting, her face looked bright.

A couple of weeks later, I received a text message from Qin: "I am on my way to Shenzhen. If I have time, I will come back and see you in Beijing. Or come and see me in Shenzhen sometime." By coincidence, I happened to be in Hong Kong, a short train ride from Shenzhen, having just attended a friend's wedding. I decided to visit Qin the next day.

The southern border town of Shenzhen had transformed itself in two decades from a small fishing village into a modern metropolis. It was one of the first areas the government had allowed to develop a market economy. Along with capitalism had come many social ills; crime and prostitution ran rampant. When I had visited a few years before, I was greeted by aggressive touts and swindling taxi drivers. But this time, after clearing immigration and customs, I was surprised to find myself on a peaceful new promenade lined with tall sparkling buildings, a shiny entrance to a new subway, and a giant billboard advertising Yahoo.

Qin and I met outside a fancy hotel. She had arrived in the city less than twenty-four hours before and was wearing jeans, a turtleneck, and a sweatshirt that were too hot for Shenzhen's subtropical weather. Her hair was pulled back in a banana clip, and when she saw me, the click of her black heels quickened on the pavement.

Qin explained that she had decided on the spur of the moment to leave Beijing. The new job was too demanding, and she had been lonely without her classmates. And Zhu was waiting for her in Shenzhen, she said, indicating a chubby young man with a pale face and a barrel chest who was approaching us, accompanied by another man. Zhu and Qin had gone to junior high school together, and their parents were friends. He had served in the army for two years, and had been posted to Xinjiang, a far northwestern province near central Asia with a large population of ethnic Muslims still wary of Chinese rule. After his discharge, he had moved back to Shenzhen, and when Qin had gone home for Chinese New Year, she had bumped into him.

"Is he your boyfriend?" I asked quietly.

She hesitated for a moment and then said, "Well, he has to prove that he can take care of me."

Zhu's friend was another Sichuanese named Little Fu. The two worked part-time as truckers, hauling computer parts, purses, and clothes made in Shenzhen's factories to transportation hubs. They lived on the outskirts of the city, and they weren't familiar with the downtown area where we were. We awkwardly stood in front of the hotel for a while, trying to decide what to do.

"What do you guys do for fun?" I asked at last.

"We get a room and sing," said Zhu. He meant karaoke.

We decided to take a taxi back to the neighborhood where Zhu and Little Fu lived. As we rode, the buildings became grayer, duller, more rundown. We turned onto Lotus Road, which was full of potholes, puddles, and trash. Dazed-looking women walked around the neighborhood wearing puffy short skirts, boots, and sagging pantyhose. Just beyond the façade that Shenzhen had erected at the border, the city I remembered lived on.

We went to a restaurant, where Qin rinsed the plates with hot tea before she allowed us to eat off them. She ordered her favorite Sichuanese dish, cat's blood bounty, which she shared with Zhu and Little Fu. I ordered a few baskets of Cantonese dim sum. As we began our meal, Qin ribbed Zhu in a Chinese way.

"He's so fat."

"He's not that fat," I said. "At least, not by American standards."

"Does everyone in America have a gun?" Zhu asked. After I answered no, he had more questions. Was there racism? What was the weather like? Did the men wear jeans? Was it true that all the men had curly hair?

I told him that no, not all the men had curly hair. Why didn't he ask the same about the women?

"Well, I noticed you don't have curly hair," he said and paused. "I hear the men are wild."

"Wild?" I asked.

"Yeah, like ready to fight."

He continued to fire away. "I hear that you can make a lot of money in America just washing dishes. How much does a bowl of rice porridge cost in America? Three dollars? Really? You can eat twenty bowls in China for that price . . ."

I had to return to Hong Kong that afternoon to catch my flight back to Beijing. On the way, I couldn't help worrying about Qin. She hadn't told her mother that she had arrived in this alien southern town. It was Qin's first time in Shenzhen, though her mother had lived there for more than a decade. The dispersed family reunited once a year, during Chinese New Year, in their hometown in Sichuan, China's heartland. When Qin had returned to Sichuan for the holiday a few months before, she had spent $200, most of her savings, on New Year's gifts—cigarettes, liquor, candy, and vitamins, for her friends, neighbors, and relatives. "I'll tell my mother I'm here when things are a bit more stable. She wouldn't be happy to know that I left my job to come down here for a boy." The thought worried me too. Zhu seemed nice enough, and Qin assured me that Zhu would help her find a decent job. "He won't let me work in a factory," she said. "He has a lot of connections." But I was concerned with all the abrupt changes in her life. And what if she got pregnant?

Just before we said goodbye, out of Zhu's earshot, I had blurted out, "Are you guys using condoms?"

"What do you mean?" said Qin, looking genuinely confused.

Then she blushed. "Oh, no. We're not doing that. We're going to work out what our relationship is first."

I cleared my throat. "Well, if you decide to do it, use a condom, okay?" It was the first time I'd ever given a lecture like that, and it was doubly hard to say it in Chinese.

It turned out to be difficult for Qin to find a restaurant job in Shenzhen because many employers demanded that new hires fork over a deposit to ensure that they wouldn't quit abruptly. Qin didn't have the money for that.

She text-messaged me one afternoon with a phone number. "Ask for number two," she wrote. When I called, I found out that it was a barbershop. Qin was making $120 a month washing men's hair. She started working at three in the afternoon and finished probably long after I had gone to bed, she said. It was very loud in the shop; I could hear someone screaming in Sichuanese.

"Do you want to come back to Beijing?" I asked.

"Yes, but I can't until I have more money," she said. She mentioned that she and Zhu had been fighting. "I prepare his baths for him and he still criticizes me," she said. "But at least he doesn't beat me."

Before we got off the phone, she said, "Take care of yourself." All I could think to do was to tell her the same.

Shortly before Zhang's wife and son were to return to Shanxi, I visited the family at their home in the Xiao Wu Ji area, on the southeastern outskirts of Beijing. It took them half an hour to bike between home and the noodle shop. The area had been farmland until a few years ago, but it was now a warehouse district.

Zhang couldn't give me an exact address because there wasn't one. A taxi dropped me at the bridge where he told me to meet him; the area looked dark and barren. I didn't see Zhang anywhere. A few people loitered at a bus stop next to the highway. The driver was kind enough to let me wait in his cab.

"This area is very disorderly," he said. "Too many migrants. Be careful. Do you know this person you are meeting?"

A few moments later, Zhang pulled up on his bicycle. He thanked the driver for waiting. As the taxi sped off, Zhang patted the rack behind his bicycle seat. "Want a ride? It will take too long to walk," he said.

I climbed on, sitting sidesaddle. As Zhang picked up speed, I did my best to balance my legs, dangling off on one side, against my head and torso, leaning out on the other. We rode through a lively neighborhood. Billiard players cued up on a series of outdoor tables. Diners ate on plastic tables set up outside restaurants that advertised all kinds of Chinese cuisine: SICHUAN SMALL SNACKS, HARBIN SPECIALTIES, and SHANXI DUMPLINGS, the signs read.

"Did you see that place?" Zhang asked as we zoomed past the last sign. "Lots of customers. There definitely is a market for Shanxi food in Beijing."

I grew tense as we turned onto a dark, wide street. Zhang kept to the shoulder, but still I cringed as big blue trucks edged alongside, shifted gears, and passed us, kicking up a dust cloud. I laced my fingers around Zhang's waist. We went by a row of warehouses with the sign BEIJING MARINE SHIPPING CO. Zhang explained that this was a place where imported goods were inspected before being sent around the country.

We turned onto a dirt road and continued in complete darkness for a minute before I could make out faint lights coming from a long, narrow building that was divided into a series of makeshift dwellings that reminded me of self-storage centers in the United States. Zhang's wife emerged from one of the units, no bigger than a one-car garage. In place of a garage door, each unit had a front door and a window.

Inside, a dim light bulb hung over a coal oven, a sink, and a hard double bed, which Zhang and Yao shared with their son. Zhang had added a lone personal touch, a framed photo of a Buddhist statue of the Goddess of Mercy. As Yao had complained, the room was hot, though it had cooled outside. I now understood why the family lingered at the canteen until late in the evening, long after closing. It was far better than hanging around here.

Zhang lit a coil of incense to scatter the mosquitoes. They didn't know their neighbors, he said, because like themselves, everyone was a migrant in transition, staying for a few weeks to a few months before they uprooted themselves for a job elsewhere.

Yao was lying on the bed with her eyes closed. "She has leg problems," Zhang said. "If she stands for too long, her leg swells up. There is something wrong with her *qi*" — her energy.

Erzi, on the other half of the bed, opened a chessboard and arranged the pieces. I played with him for a while before I left.

A few days later, Zhang's wife and son packed a few belongings and returned to Shanxi.

I visited Zhang next on his last day in the canteen, which happened to be his hundredth day in business there. He wasn't sure what he was going to do after closing the shop, but he decided it didn't make sense to keep it open if he was barely breaking even.

Like Zhang and Qin, I was also on the move. In a few short weeks I would be heading to Shanghai to intern in an upscale restaurant on the city's riverfront. When I told Zhang that dinner there could run upward of $100 per person, he asked with a laugh, "Who are they trying to cheat? Rich Chinese or foreigners?"

He was almost done cleaning up the kitchen. The refrigerator was empty, and only a few *jin* of flour was left.

"This is your last chance to eat my noodles," Zhang kept repeating to the regulars who streamed in at lunchtime. He brought a bowl of noodles with tomato-and-egg sauce to a woman who wore the same green gardening apron every day.

"Where are you going?" she asked.

"Don't know. Do you want your noodles for here or to go? It's your last day to eat my noodles. You might as well eat them here."

She took her bowl and sat down in the mess hall.

"If it's not salty enough I can add more sauce," Zhang called to her, but she seemed to have fallen into a trance as she slurped.

Another regular asked for his usual, a bowl of knife-grated noodles with pork sauce. I offered to grate the noodles, but Zhang shook his head. "He needs them really thin. He has health problems," Zhang said, pointing to his diaphragm.

Near the end of the rush, he sat down on a stool behind the counter to rest his feet. The black canvas shoes he wore didn't give much support. Dark shadows circled his eyes. His blue veins protruded from his bony arms and hands. His wrists were swollen.

"I don't have any regrets," Zhang said. "The days went by fast."

He resolved to press on. He'd find a new space, maybe with Old Wang, maybe on his own. He was determined to succeed. Whatever happened, he would never go back to Yushan. "Why work for the Communist Party? I should work for myself," he said.

When the last customer finished his noodles and came to the counter to pay, Zhang smiled like a proud mother. "You finished the whole bowl, didn't you?" he asked. After he collected and counted the money, he turned off the lights in the kitchen and brought a bowl of his noodles into the mess hall. He hunched over and slurped until they were all gone.

◎ Side Dish 2: The Rice Harvest

The guidebooks informed me that the jagged, humped landscape of Ping'an, embedded in mountain folds, resembled a dragon's back. I thought that each peak looked more like a wedding cake, the tiers of which were rice paddies that farmers had carved into the mountains centuries ago. It was autumn, and the long stalks of rice were turning from bright green to golden yellow as they ripened.

I had come to Ping'an to harvest rice, the staple of China's south, the constant of my childhood. I was also hoping to find, in the lush terraces of relatively undeveloped Guangxi province, a respite between smoggy Beijing and the frenetic pace I knew I'd find in Shanghai. I'd flown into a touristy town called Guilin, then taken a bus that careened around hairpin turns as it climbed into the mountains, techno music pulsing through the speakers. I had to hike the last mile, regretting the decision to bring a suitcase rather than a backpack. Back in Beijing, I'd forgotten that there weren't any roads in Ping'an.

With no roads, it was impossible to bring in the machines farmers increasingly used to harvest rice elsewhere in China. In Ping'an, farmers threshed by hand, just as they had for seven hundred years. I admired those terraces from my guesthouse window for days. First rain had delayed the harvest, heavy downpours alternating with light showers. Then the government told the farmers to hold off until an important guest—Lien Chan, the chairman of the Nationalist Party in Taiwan—passed through.

Not everyone anticipated the harvest as I did. After the farmers mowed the rice stalks, Ping'an lost its sparkle, and thus its tourists. The beauty and the tourist dollars would return in the spring, with a new season of rice. It began with the farmers planting seeds and flooding the paddies, which turned the mountainsides into a mosaic of mirrors.

So the farmers waited patiently, drinking beer and *baijiu,* watching soap operas, and playing mahjong. The dawn cry of roosters set off a buzz of activity on the ground floor of my bed-and-breakfast (and lunch and dinner, as it's done in the Chinese countryside). Auntie Liao, the proprietor, chattered with a relative in rapid-fire fashion. Uncle Liao began sawing bamboo in the yard underneath my window. These sounds drifted up to my second-floor room as I lazed in bed. I didn't have my normal routines—bicycle, market, kitchen—to keep me busy.

"Are you bored yet?" Little Su, one of the housekeepers, asked me every day. I wasn't, really. It was refreshing to stare out at the staircase of terraces that wound up the mountain, the green and yellow stalks glowing like neon, and the distant peaks almost impossible to distinguish from the gray sky in the mist.

On the fifth day of my stay, the important guest passed through, and on the sixth, the sky was overcast but the rain had ceased. Today the harvest would begin.

I was eager to get out in the fields, but Uncle and Auntie Liao sat me down to a leisurely breakfast with their relatives. We squatted on stools at a large round wooden table, and the farmer to my right offered me a drink. Alcohol had already ruddied his face, and his right eye drooped. In this region, liquor came with every meal, including breakfast. He extolled the benefits of freshly distilled sour-plum *baijiu*, but I stuck to beer as the farmers filled their cups and toasted. Then we picked up our chopsticks and dug into a pot that sat on an electric hot plate in the center of the table.

Though the ingredients varied every day, the meals in the village were cooked the same way, fondue style. Today the stew was enriched with fermented tofu, chili peppers, and dried anchovies, eaten whole. The villagers chewed the tofu and fish and paused to slurp the soup, bringing their bowls to their lips. It took a while to get used to the fermented tofu, which had a springy texture and a strong, ripe flavor, not unlike aged blue cheese. *Baijiu* and stinky tofu were the Chinese farmers' version of wine and cheese.

Flushed and dizzy from the beer, I followed Uncle Liao into the endless rows of stalks. To avoid immersing our feet in the muddy water, we walked on a two-foot-high stone wall at the edge of a terrace. More accurately, Uncle Liao walked and I teetered precariously on the stones, which were barely wider than a balance beam. The alcohol was doing me no good.

We stopped at a plot that was slightly drier than the rest. The rice stalks grew in bunches as tall as my waist, the grains embedded in the yellowing tips. Uncle Liao's sister and brother-in-law were bent over, already at work with their sickles. They each grabbed hold of a bunch of stalks and freed them with swift hacks. Uncle Liao handed me a sickle that felt as heavy and awkward as a cleaver once had. I tried to imitate their motions, squatting in the paddy as the rice stalks rose like a jungle around me. I swung my sickle at the base of one clump, expecting it to come loose effortlessly. But the sickle got wedged in the stalks, and I had to saw at them to extricate the blade. More than a few times, the stalks poked me in the eye.

"Want to thresh?" Uncle Liao asked, shaking his head. He'd had a hard time understanding why I had wanted so desperately to be a part of the harvest, but he'd been nice enough to accommodate me.

Stalks that had been cut earlier in the day hung over the edges of the terraces, drying. The farmers pushed a long wooden crate along a plot already stripped of its stalks. Uncle Liao's sister, about five feet tall and no more than a hundred pounds, gathered up a bunch of stalks over her shoulder as if she were holding a baseball bat, then swung, whacking the stalks against the inside of the container. In two swings, the stalks had rained their contents into the crate: ladybugs, beetles, and grains of rice encased in brown husks.

Uncle Liao's brother-in-law watched me try. "Not bad," he said.

"You're much faster," I said.

"Not as fast as a machine!"

Uncle Liao said the average farmer could thresh three hundred pounds of rice a day—enough to feed one person for a year. In an hour, the four of us had half filled the crate, enveloped in a flurry of chaff more potent than fresh-cut grass. I began to sneeze. The back of my throat itched. My head was still pounding from the beer I'd had at breakfast. The farmers laughed.

"We all get a little allergic," Uncle Liao said. He had a remedy: "You take pig's blood and mix it with cow's blood and make it into cakes."

I went back to cutting stalks, hacking away until I reached the end of a long plot that spanned three hundred feet or so. I felt good until I looked up and saw dozens more overgrown plots with stalks reaching toward the sky. Bring on the *baijiu*, I thought.

All afternoon Uncle Liao and his brother-in-law went back and forth between the fields and the village, carrying hundred-pound sacks of threshed rice over their shoulders. The next day, so long as it didn't rain, the rice would be spread on a patch of concrete to dry. Then it would be taken to the general store and poured into the funnel of a sprawling, ancient metal contraption that spit out billows of husks and sent the grains rushing down a chute to the floor.

The rice harvest had always been important to Ping'an. In the past, rice was the villagers' livelihood, their sustenance. Villagers subsisted on a diet of rice steamed with pumpkin and taro root. Rice was currency: when they were younger, the Liaos' sons each carried seven pounds of rice to school every week, which helped pay their tuition. But the meaning of rice had shifted in the past decade. Now the rice terraces served as backdrop for a more lucrative industry—tourism. The Liaos tended their guesthouse and restaurant rather than their fields, which were overseen by poor farmers they hired from nearby villages. The Liaos and the other Ping'an farmers had also pushed back the harvest by a couple of weeks so that October tourists could see the fields before they were stripped bare. Most Chinese got the first week of October off to commemorate the founding of the People's Republic, and they celebrated in capitalist and consumerist style, by going on travel and spending binges. A good number of them flocked to Ping'an.

Uncle Liao had been a rice farmer since he'd finished middle school. In the 1980s, he left the village to work as a tea trader, traveling between Hunan, Guangdong, and the island of Hainan. In 2000, news reached him that a fire had destroyed his house, along with twenty other wooden structures in the village. He returned to Ping'an and, recognizing the area's tourism potential, rebuilt his home as a guesthouse.

Now Uncle Liao was in his forties. He had a boyish face, and the baseball cap and T-shirt he often wore made him look young. When he went about his daily tasks, he slouched like a teenager sullenly doing his chores. The Liaos' two sons, both in their twenties, lived with them, but neither helped much with the guesthouse or the rice. Auntie Liao explained that she and her husband were hesitant to burden the kids, fearing they would move away to the city.

Auntie Liao was a pretty woman with a wide smile who wore her long, frizzy hair in a braid down her back. Like nearly all the village women, she was short, making me feel like a giant at five foot four. Unlike the other women, she had given up the custom of wearing a heavy terry-cloth-like bandana, but she wore traditional, flowing shirts and pants in black with bold blue edges.

The villagers of Ping'an belonged to an ethnic minority

called the Zhuang, who had been physically isolated by the mountains for centuries. Though they didn't look so different from the Han Chinese, the dominant ethnic group that made up more than 90 percent of China, their language sounded nothing like Mandarin. The villagers told me the Zhuang language was related to Thai, although it had no written form. The language was being replaced by Mandarin, which children learned in schools and adults picked up from the tourists. But many of the elders could communicate only in Zhuang.

Ping'an—population 800—was like a giant family; everyone had the same surname, Liao. Uncle and Auntie Liao, like many of the village's couples, were both born in Ping'an but came from different tribes, which ensured that close relatives did not marry.

Until the 1970s, the nearest road was a three-hour walk away. In the 1980s, the government paved a two-lane highway that stopped an hour away, and in 1997, when Beijing designated the village a tourist development zone, the road was extended, snaking up the steep hills to a parking lot a twenty-minute walk from the village. That was where the bus had left me, and like other visitors to Ping'an, I'd been charged admission. When the bus had paused at a gate just below the mountain, a ticket seller had hopped on and collected $6 from everyone who looked like a tourist. I was told to hang on to my ticket for the duration of my stay.

The Liaos' guesthouse was built into the mountainside, and they ran a small restaurant just down the slope. Every morning, they opened the restaurant's hinged wall, which faced the village's main pathway, exposing the restaurant to passersby. It was like eating in a diorama. The dining room was suspended on a platform that jutted from the mountain-side. Narrow gaps between the long wooden floorboards revealed a steep drop, and I tried not to think about the shoddiness of most Chinese construction as I walked on the creaky boards.

The kitchen, which the Liaos allowed me to use, looked out onto a forest of bamboo and pear trees, with rice paddies in the distance. Unfortunately, the peaceful atmosphere was often ruined by the proprietors' attempt to compensate for what they thought of as a backwater setting by amplifying off-key karaoke singing so loudly it reverberated through the valley.

The dishes the Liaos served were simple, stir-fried combinations of local ingredients. Aside from rice, the family grew tomatoes, sweet potatoes, chili peppers, and corn. The produce was bursting with flavor, as I discovered one morning when I tasted tomatoes with a tangy, sweet flavor that was as strong as the garlic I had stir-fried them in.

"Our tomatoes are good because of the water and because we don't use pesticides," Auntie Liao said. "They may not look pretty. Sometimes they have bugs. But they taste good."

Rice, of course, was also a specialty. My favorite dish was a mixture of glutinous rice, called *luomi*, slices of sausage, and bits of taro, plus a couple of spoonfuls of water, oyster sauce, and soy sauce. All this was stuffed into a bamboo tube, which Uncle Liao had chopped into one-foot sections, and sealed with a corncob on each end. The tubes were roasted for half an hour over an open fire. When the outside became charred, the tubes were split open with a knife. The rice — each tube a generous serving for one — was moist, sticky, and flavorful.

The electric hot pot was the Liaos' prized possession. After locking up the restaurant for the night, Auntie and Uncle Liao went back to the guesthouse, a dozen yards up the mountain, and set a pot on a hot plate on the dining table. Once the soup boiled, everyone dug in with their chopsticks. Relatives soon dropped by. More food was added to the pot, and the heat was raised so that the soup again bubbled and boiled. Villagers could go from house to house, eating free meals for months that way. The soup always contained a base of wild mushrooms, bamboo shoots, ginger, and scallions, but the meats varied. Sometimes it was bull frog. Other times it was pork trotter, the giant rubbery feet giving the soup a greasy gleam. Now and then a member of the family scooped up an unsuspecting free-range chicken that strutted around the village.

I watched Daixun, the Liaos' younger son, slaughter a chicken one evening. He grabbed it firmly in his hands, ripped a patch of feathers off its neck, pulled its head back as if it were made of rubber, and slit its throat with a cleaver. He turned the chicken upside down, held up one of its legs, and drained the dark red blood into a bucket. The chicken kicked with its free leg for a few minutes until all the blood oozed out. Then Daixun put the animal in a washbasin, poured boiling water over it, and plucked the feathers. This was the first time I had watched a living thing other than a fish being killed, but it did not unsettle me. It was an easy death.

Auntie Liao usually stir-fried a couple of dishes as well, but the hot pot was used every day. It was the Zhuang version of a microwave oven—a new appliance that simplified cooking and so was instantly overused. It also made the food less interesting. I much preferred Auntie Liao's stir-fried dishes, but she said, "Hot pot is easy, and you can reheat it easily if it gets cold. Stir-frying is more trouble. We don't like to be too far from our soup." During my visit, the hot pot failed to appear only once, when the village's electricity went out.

Though rice was all around them, the Liaos ate little of it, preferring meat of any kind. I quickly discovered that it was an imposition to ask for rice at dinner. "Uh, let me see what I can do," said Little Su, but she came back empty-handed. The rice was locked away in the restaurant for the customers.

The Liaos always waited for me to come back from my afternoon hike before they began dinner. They expected me at every lunch, too, which I discovered only after I returned from an American-style brunch at one of the village's few Western restaurants. I had grown tired of hot pot, and with all the relatives dropping in, I figured they wouldn't miss me. But when I returned, Uncle Liao, Auntie Liao, and their sons took turns interrogating me. What had I eaten? Was it good? If I didn't care for their food, I should tell them, and they would make something else.

RICE VERMICELLI WITH TOMATOES
(*FENSI CHAO FANQIE*)

2   tablespoons vegetable oil
1   teaspoon minced ginger
2   teaspoons minced garlic
3   large tomatoes, diced
1   cup winter bamboo shoot, diced
    (preferably fresh)
3   fresh shiitake mushrooms (or dried, soaked in
    water), thinly sliced

2 bunches rice vermicelli
¼ teaspoon salt
1 tablespoon soy sauce
1 tablespoon oyster sauce

Place the oil in a wok over high heat for 30 seconds. Add the ginger and garlic and stir for 1 minute. Add the tomatoes, bamboo shoot, and mushrooms and stir for 3 minutes. Reduce the heat to medium and cook for 3 minutes. In a separate pot, boil the vermicelli in water for 1 minute and drain. Add the salt, soy sauce, and oyster sauce to the wok, then add the vermicelli. Toss, remove from heat, and serve immediately.

What I couldn't get used to in Ping'an was the constant flow of alcohol. At every meal, I did my best to deflect offers of beer, *baijiu,* and whatever imaginative cocktail my host made from *baijiu,* only to end the meal red-faced and inebriated. First thing in the morning, Auntie Liao cracked open bottles of beer the way mothers set out cartons of milk.

"We didn't use to drink this much, just on special occasions," she said, her tanned face flushed pink. But as their lives improved, so did their tolerance. "It's my habit now. I have to drink every day."

Uncle and Auntie Liao invited me to join them for lunch at a relative's house one afternoon. It was two o'clock, well past the time that most Han Chinese ate lunch. A dozen elders and middle-aged folk sat around a big table, eating and cackling. They were celebrating the visit of a young man who had returned to Ping'an for the first time in seven years.

That called for a lot of drinking.

One of Uncle Liao's sisters, who sported a yellow bandana,

hoop earrings, and a gold tooth, went around with a bottle of *baijiu* as if she were putting out fires in people's cups. Everyone spoke in rollicking Zhuang and laughed in unison. "Woooo! Hahahah! Wahahaha!" A relative whom everyone called the Prince—he was wearing a white dress suit— chugged two glasses of *baijiu*. When he saw the concern on my face, he said, "Don't worry! We're having fun!" before he staggered out the door and disappeared.

The visiting man sat uncomfortably at the table with his wife and toddler, looking as if he wanted to make himself vanish. I was uncomfortable too. I concentrated on the chicken hot pot, trying to improve my chicken-eating skills. Everyone else seemed to be able to insert a piece of bony chicken into their mouths, chew and crunch, spit out the bones, and swallow the meat effortlessly. At least in the midst of all the drinking and merriment, no one would comment on my ineptitude.

Across the table, a mustached relative wearing a China Mobile T-shirt and a cell phone on his belt was hunched over in his chair trying, with great effort, to hold his head up. His glass contained *baijiu* and Future Cola, a Chinese knock- off of Coca-Cola that was even more chemical tasting than the original drink. After downing a glass, he glanced in my direction and proclaimed, "Drinking is entertainment!"

China Mobile was the first to notice that the yellow- bandana lady had been filling everyone's glass but wasn't drinking much herself. "*You* should drink!" he shouted. A scuffle ensued as he tried to pour a glass of *baijiu* into her mouth. She clamped her mouth shut and the liquid dribbled down the front of her shirt. China Mobile grunted and yelled, and then, holding an unlit cigarette between his middle and ring finger, staggered and fell to the floor.

"Does this happen often?" I asked Uncle Liao.

"No. We do this much drinking only when relatives visit," he said cheerfully. Meanwhile, the visiting relatives had moved away from the table, away from the drunkenness, and onto the sofa, where they kept their gaze on the television. I finished my beer and decided to leave before anyone forced me to drink more. When I excused myself, Uncle Liao said, "Okay. I'll bring them over to the house tonight."

When dinnertime rolled around, everyone reassembled, except for two victims the lunch had claimed: the Prince and China Mobile. I sat next to the woman with the yellow bandana, who slumped in her chair and had bloodshot eyes. Beer was poured, but she pushed her glass away. "I'm still drunk," she said. Uncle Liao took pity on her and gave her a cup of tea and a glass of Coca-Cola. This time, it was the real stuff.

At a more sober meal, I quizzed Auntie Liao on how much rice she used to eat, before tourism came to Ping'an.

"We used to eat two bowls of rice a day," she said.

"Two bowls?" howled a man at the table. "More like five bowls!" A few whiskers decorated his chin, and his ears protruded from his bald, egg-shaped head. He wore a faded blue Mao jacket and dark pants. Since the Liaos never bothered with the formality of introducing me to anyone who ate with us, I assumed he was just another relative from the village. But he turned out to be a farmer who worked in the Liaos' fields. He was from a village about thirty miles south of

Ping'an that was less picturesque and therefore poorer than Ping'an, because it couldn't attract tourists. But its lower elevation supported two rice crops a year, and it was accessible enough that the threshing could be done by machine. "The first crop takes longer to grow because it's earlier in the year, which means the water is colder," he told me. "It's like cooking rice in a rice cooker. If you start out with cooler water, it takes longer to cook."

The farmer's last name was Long, meaning "dragon," a common surname. When he told me his first name, I asked him to write it down to make sure I had gotten it right. He fumbled with the pen, and it occurred to me that I might have finally met someone who was more illiterate than myself.

"My first name is Yuntu," he said. "*Yun* as in 'moves,' and *tu* as in 'dirt.'"

"Your name is Moves Dirt?"

"Yes. Moves Dirt. My parents wanted the word 'dirt' in my name because they wanted something that referred to the land reforms of Mao's time. They admired what Mao did for the farmers," he said.

Dragon Moves Dirt. It was a fitting name. In the lulls between planting his rice and harvesting it, Farmer Dragon went up the mountain to work for people like the Liaos, who had become relatively well-off from tourism and paid him about $5 a day for his labor. Once he was done with the rice harvest in Ping'an, he would head to a slightly higher elevation and thresh rice there. When the rice season was over, he picked mandarin oranges down in the valley.

The next evening, I ran into Farmer Dragon near the

guesthouse. It had rained that day, so there was no threshing to do. He had spent the day "playing," he said.

"Playing what?" I asked. "Mahjong?"

"No, I don't enjoy that sort of thing," he said.

I realized he simply meant that he hadn't been working. He had been hiking, he said, and was now on his way back to where he was staying. I joined him on his walk.

We made our way down a stone path that wrapped around the mountain until we reached an abandoned farmhouse on the outskirts of the village. It was dark inside, save for the flicker of a fire in a cement pit. This was where he cooked. Through the window, I could barely make out the contours of the rice terraces in the dusk. Only when they faded into darkness did he flip a switch that turned on a couple of dim bulbs.

Sawdust covered the farmhouse floor. Aside from an incongruously elegant grandfather clock and a few stools, there was no furniture. The abandoned house belonged to a villager with many homes who allowed migrant farmers to turn it into a hostel, Farmer Dragon told me.

Farmer Dragon chatted about his background. Like me, he was Han Chinese, and Mandarin was his mother tongue, which made communicating with him easier than it was with some of the Ping'an villagers. Farming was all he had ever known. He had a fifth-grade education, if it could be called that at all. The school held classes three days a week; the rest of the time the teachers made the students do manual labor.

"We plowed, raked, and planted. We paved roads. We worked from seven in the morning until seven at night, with a two-hour break at noon. It was right after Liberation, and our country was very poor. We had to start somewhere. I finished fifth grade when I was sixteen.

"Back then, everyone was afraid of being labeled a capitalist. I remember my mother at one point had raised seven ducks—six females and one male. The village leaders told her to get rid of two of the ducks, because each household was only allowed to have five."

As he spoke, a pot of rice simmered over the fire. A fellow migrant sat nearby, cutting watermelon rinds. Was that what they were going to eat? I wanted to hear more about his life, but I didn't want them to feel obligated to share what looked like meager rations. My sentiments weren't entirely charitable; it didn't look very appetizing either.

"I should go," I said. The Liaos would be waiting for me, I told him, and I was relieved when Farmer Dragon didn't protest. We agreed to meet again the next day, which would be my last in Ping'an. He promised to take me threshing in the fields if the weather was good.

The sun was shining when Farmer Dragon and I met the next morning, outside a guesthouse around the corner from mine. As we waited for the owner to show up to give Farmer Dragon his day's assignment, a thick fog rolled in, obscuring the rice paddies and dimming the possibility of work in the fields.

"What happens when farmers don't have work in America?" he asked. "Do they do something else?" I explained that most of the agricultural work in my native state of California was done by Mexican workers, who crossed the border for better opportunities. If they didn't do farm work, they often got jobs in restaurants or as domestics.

He nodded. The Mexicans were like him, he said, migrating where there was work to be done.

We waited another ten minutes for the weather to clear, then he said, "Let's take a walk instead."

He proposed that we go to Middle Six, the next village over, which was inhabited by an ethnic minority called the Yao. Yao women who followed tradition didn't cut their hair, instead wrapping the long strands around their heads like birds' nests. I had seen them in Ping'an, trying to profit from the tourist boom by unfurling their freakish, Rapunzel-like tresses and demanding money from anyone naïve enough to snap a photo.

We followed a muddy path that ran along a ravine. I was an anxious hiker, afraid of snakes, getting lost, falling, and running out of water. In some places it was steep and narrow, and Farmer Dragon scrambled up the inclines with the agility of a mountain goat. On the flatter sections he paused, pulled out a plastic bag of tobacco, rolled a cigarette, and smoked it as he strode along. I heaved up the trail behind him.

Farmer Dragon turned to look back, his gaze falling on my expensive hiking boots. "Those shoes are hard, so you have less grip. You'll fall easily," he said. "Mine are more flexible." He was wearing cloth shoes that didn't cover his ankles. He continued to talk in his calming way. "When there's a fork in the road, always take the wider path. The wide path will always get us where we want to go."

At a set of benches under a straw canopy, we rested while he rolled another cigarette. "I know this path well," he said. "I used to walk along here at night. My girlfriend used to live in Middle Six, so after a day in the fields in Ping'an I would go to see her. This is where I'd stop."

Farmer Dragon was a widower. He said, "My wife was always sick, and she had been sick for three years before she died. We didn't have the money to go to the hospital to find out what was wrong with her. Even when I first met her, she didn't look healthy. She didn't look womanly. It was an arranged marriage. I met her twice, both times with her parents present, and then we married. My family gave a dowry of a hundred dollars. That was a lot of money. Back then, pork was only thirty cents a pound!"

After his wife's death, Farmer Dragon was introduced to a woman through a friend. The woman had recently separated from her husband, and she and Farmer Dragon got along well. "I came to visit her every so often and spent many nights there when I was working in the mountains," he said. But there had been a complication.

After he and the new woman had dated for a year or so, "she moved back in with her husband down in the valley. She was Yao. Yao women are known for being loose. They have casual relationships. She decided she couldn't leave her husband. She had to think about her family obligations. She had two children. And who was going to take care of her husband's parents? So now she comes to visit me two or three times a year."

He didn't sound bitter. I asked, "It doesn't bother you that she's with someone else?"

"I'm okay with the situation. She comes when she wants to, and we enjoy each other's company. We're both old. We don't have many expectations."

He fell silent as we entered a valley with a stream, from which he scooped water and drank from his hands.

"Do you believe there are ghosts in this world?" he asked.

"I saw one here once. It was dark. I was sitting where we were just sitting and I heard a loud crash. I couldn't see anything, but I knew it was a ghost."

"Couldn't it have been a person?"

"It was definitely a ghost. Nobody walks here at night. And some people, when they die, aren't satisfied with how their lives turned out. A lot of people died in ways they shouldn't have, especially during the Cultural Revolution. Those people stick around."

I was glad when we reached Middle Six, a small cluster of wooden houses by a river that ran through sloping rice paddies. A light rain was falling. Several villagers staring out a window on the second floor of a farmhouse recognized Farmer Dragon and invited him in for a meal.

Inside, a Yao woman was combing a wig of knee-length hair.

"What is that for?" I asked. I felt as if I had caught Sleeping Beauty in her Disneyland dressing room.

"It's not my hair, but it's real. Adding more hair makes me prettier," she said, smoothing down her tresses. She was preparing to hike to Ping'an, to work the tourists for the afternoon.

"Stay for lunch," said the woman, putting on her wig. "Don't worry, we won't charge you."

I told Farmer Dragon that I had to get back to Ping'an for a final meal with the Liaos. I asked for his phone number, but he said he didn't have one. He wrote down his address, minus the postal code, which he couldn't recall. We exchanged an awkward goodbye, then I hurried off before I remembered to take his picture.

part three

FINE DINING

11

My mother visited me in the spring of 2003, when I was living in Shanghai and had yet to start my cooking adventures. We spent a week traveling around China, visiting the historic sites she had learned about in elementary school in Taiwan and eating in grungy canteens, quaint teahouses, and the occasional opulent dining room.

The week passed quickly, and before I knew it we were back in Shanghai. Waiting for the car that would take her to the airport, I felt sad that the time had come to say goodbye. My mother was feeling sentimental too. She dashed out of my apartment, leaving behind the suitcase that she had packed. Before I could figure out where she had gone, she returned with a Styrofoam container. In it were half a dozen *xiao long bao* —round, palm-sized dumplings with folds that resembled a belly button's. They were plumped up like water balloons around a pork stuffing and a spoonful of hearty soup. In our final moments together, she ate the dumplings with a single-minded focus, until the last juicy drop was consumed and all that remained was the faint buttery aroma of the broth.

The dumplings meant more to my mother than seeing the vast maze of the Forbidden City and the crumbling watch-towers of the Great Wall. "I had heard about *xiao long bao* as a child," she later told me. "But it wasn't until I went to China that I had the real thing." The Shanghai delicacy was the highlight of her trip.

Three years later, in the fall of 2006, I traded in my noodle knife for a starched chef's jacket and checkered pants at a top-end restaurant in Shanghai. I was back in the city where I had first discovered my love of eating.

Shanghai had an allure, a glamour, that Beijing lacked. If Beijing could be personified as a brusque, burly man, then Shanghai was his manicured female cousin. Where Beijing's streets were wide, orderly avenues that formed 90-degree angles, Shanghai's were narrow and curvy. Plane trees like those in the parks of London and Paris canopied the roads. As in Beijing, development ran rampant, but Shanghai's towers were bunched closer together, giving the city a defined skyline that featured the space-age pink baubles and spire of the Oriental Pearl TV Tower and the eighty-eight-story, eight-sided Jin Mao building, which would soon relinquish its title of tallest structure in China to a neighboring skyscraper.

The city had been built largely by the British, the French, and other colonial powers in the late 1800s and early 1900s, when it earned the nicknames "Paris of the East" and "Whore of the Orient" — epithets that perpetuated its image as a flashy, hedonistic metropolis. While some of the European architecture had been demolished in Shanghai's drive to modernize, a number of French villas with wrought-iron balconies still stood in the middle of the commercial district and stately British limestone buildings lined the riverfront. And the city still featured *shikumen,* traditional three-story row houses, unique to

Shanghai, which combined Chinese elements, like courtyards and high gates adorned with ornate stone carvings, with familiar Western features, like fireplaces, shutters, and small attic windows poking from the red-tile roofs.

I had learned the geography of the city by eating my way through it. I knew the main shopping artery of Nanjing Road for the crowded shop where patrons elbowed each other to score a bamboo basket of *xiao long bao;* dirty, bustling Wujiang Road as home to a narrow row house that served dumplings grilled in a skillet the size of a wagon wheel; the elevated Yan'an Expressway for the man at the takeout window beneath who served Shanghai *shaomai,* a loosely wrapped steamed dumpling with a ruptured top that resembled a volcano. And then there were a string of nameless alleys I knew only by their aromas, like the lane behind a nightclub where a man in a Muslim skull-cap grilled cumin-scented lamb skewers long past midnight.

As in Beijing, eating out was so cheap and convenient that I never had to cook at home, and there were always friends who wanted to go out for a meal. When I started writing about food, toward the end of my three years in Shanghai, I discovered that discerning diners didn't consider much of what I had eaten in restaurants to be authentic Shanghainese at all. The Shanghai soup dumplings that my mother and I loved, for example, weren't originally Shanghainese, locals told me. In fact, Chinese held Shanghainese cooking in such low regard that it didn't make the list of China's top eight official regional cuisines —a list that expanded upon the "four big" cuisines I had learned in cooking school. How was it, I wondered, that one of China's most important cities didn't have a definable, developed cuisine of its own?

The reason, I learned, had to do with Shanghai's status as a latecomer in a country with a five-thousand-year history. As regional cuisines thrived in various parts of China through the Ming and Qing dynasties (roughly from the fourteenth to the early twentieth centuries), Shanghai remained a poor fishing village that emperors bypassed en route to Yangzhou, Hangzhou, and Suzhou, whose most esteemed chefs they summoned to the Forbidden City to give the northern elite a taste of exotic cuisines. With a population of half a million in the mid-1800s (as compared to 18 million today), Shanghai didn't have a single proper restaurant, only little canteens that catered to fishermen working the Yangtze River Delta.

The area around Shanghai did develop a few distinctive culinary characteristics. Thanks to the technique of red-braising, classic Shanghainese dishes were described as *nong, you, ci,* and *jiang*—that is, thick, oily, red, and soy-sauced. Cuisine described as thick and oily didn't sound appealing to me, but in a country familiar with famine, the red-braising technique caught on in many provinces. (Chairman Mao, who hailed from Hunan, considered red-braised pork one of his favorite dishes.) The technique itself was simple: an ingredient like pork belly, fish, or eggplant was stir-fried in a mixture of oil, sugar, and soy sauce, which was then reduced until the sauce caramelized. A gooey mess resulted that could be as deliciously rich as hot fudge or as cloying as an oil slick.

Shanghai widened its culinary repertoire after China lost the First Opium War. As a term of the peace treaty, China ceded the riverfront to the British. Soon after, the United States, France, and Japan also built settlements there, and by 1870, Shanghai was the world's seventh-largest port. By the 1930s, Shanghai was among the world's most cosmopolitan cities. Foreign visitors poured in—among them Charlie Chaplin and Albert Einstein—and chefs flocked there from all over China. The city culled seafood from the nearby Yangtze River Delta and more exotic ingredients from the ships passing through its port.

"Ships calling at Shanghai regularly brought lamb and butter from New Zealand, beef and citrus from Australia," the journalist Irene Corbally Kuhn wrote of 1930s Shanghai. "The Dollar Line ships from San Francisco and Honolulu were an occasional source of expensive fresh green vegetables bought from the ship's stores." The city borrowed cooking methods, too, from once great neighboring cities, which saw their influence diminish as Shanghai rose. Cooks began stir-frying with rice wine from the nearby town of Shaoxing. From Yangzhou came the knife techniques that could turn a piece of tofu into a heap of fine shreds. From a nearby village came the *xiao long bao*. And the chefs brought the influences of faraway provinces known for their food: Canton, Sichuan, Hunan. Restaurants figured prominently in the new Shanghai lifestyle. "We had a dazzling array of choices," Kuhn observed, "for restaurants abounded and ranged from the elegant formality of the St. Petersburg, owned and managed by a White Russian cavalry officer, to the small, dark, and steamy noodle shops of the old Chinese walled city."

The new cuisine that resulted from Shanghai's sudden wealth and outside influences earned a new name—*benbang*, which can be loosely translated as "nouvelle cuisine." It was considered a bastardized form, something akin to today's fusion cooking, both admired and despised, and still excluded from the list of China's regional cuisines. (Today restaurants in Shanghai with foreign names like Paul's and Jesse serve delicious *benbang* dishes.)

In the Communist years, eating culture declined in Shang-

hai, like everywhere else in China. Restaurants were nationalized, shipping lines with the West were severed, and the socialist government randomly slotted people into cooking jobs like
so many pegs in a board. Stan Sesser, a food writer and one of
the early foreign visitors to Shanghai after China opened up
in the late 1970s, wrote that the food there was "literally slop,
slapped in front of you in drafty hotel banquet halls or in
equally dismal state-owned restaurants. The meat was gray, the
sauces grayish-brown, and the staff glared at you in open re-
sentment that you were forcing them to work."

After decades of culinary stagnation, Shanghainese food
began moving forward again after economic reforms took hold
in the 1980s and 1990s. By the time I reached Shanghai, international influence had surpassed what Kuhn had described.
Ships were bringing not only beef from Australia and lamb
from New Zealand but also *yuzu* (a citrus fruit) from Japan and
truffles from France. Chefs were coming not just from all over
China but from all over the world to reinvent a cuisine that
had been invented just a century before.

At the forefront of change in Shanghainese cuisine was a new
restaurant on the Bund called the Whampoa Club. Up until
the opening of the Whampoa Club and a slew of other high-
end establishments that followed, the district of British limestone buildings on the riverfront had been curiously lacking in
eateries, or any life, for decades. Under British rule, the Bund
buzzed with bankers, diplomats, and expatriates, but after the
Communist takeover, international businesses on the Bund had
been shuttered and the buildings left to decay under a regime
that had little interest in preserving symbols of colonialism.

That began to change during the time I lived in Shanghai,

as real estate prices skyrocketed and the increasingly mercenary government saw an opportunity to lease state-owned land
for large sums. Financed by a group of overseas investors, one
British neoclassical building, dating from 1916 and with arched
windows and doorways, was renovated in the early 2000s into
a hub of luxury brands. Called Three on the Bund, it contained
an Evian spa, a Giorgio Armani flagship store, and a restaurant owned by a French chef named Jean-Georges Vongerichten. The Whampoa Club, on the fifth floor, served Shanghainese dishes, but had taken the concept of "reinvented"
cuisine one step further than most of the city's restaurants by
using Western-style presentation and foreign ingredients and
creating a glorified colonial atmosphere on the Bund — concepts that, at the time, were considered revolutionary.

The Whampoa Club's interior unabashedly evoked 1920s
Shanghai, an era romanticized by the West but denigrated by
Chairman Mao for its hedonism. The lobby featured an antique rocking chair and an animal-print rug; wooden lamps in
kaleidoscopic patterns dangled from the ceilings. Speakers
played piped-in Shanghainese jazz recordings popular in "pre-
Liberation" times. The restaurant's windows looked out over
the Huangpu River, the artery that had made Shanghai one of
Britain's most successful ports. That the Whampoa Club was
decorated in a way that celebrated Shanghai's colonial past was
a sign of the changing times.

Leading this new restaurant was Jereme Leung, a *huaqiao,* or
overseas Chinese, who arrived in Shanghai from Singapore in
2003. Jereme (pronounced like "Jeremy") was a short, compact
man with a military-style buzz cut that reflected the regimented
routines of an exacting chef. His baby face still produced the
occasional pimple. His eyelids creased into a straight, narrow

line above his pupils. He joked that his epicanthic eyelids, a common trait among East Asians, helped keep him awake for more hours of the day.

Jereme struck me as a youthful, eager chef when I met him on one of the restaurant's first days in 2003. He was thirty-two years old, a young age to be heading a Chinese restaurant. I was writing an article about the culinary renaissance of Shanghai for *Newsweek,* and Jereme invited me to try his food. I had never had Shanghainese cuisine—or any Chinese food for that matter—so refined, carefully planned, and well executed. Particularly memorable was a dish that featured foie gras, which was just being introduced to Chinese restaurants. The pan-fried goose liver sat on a bed of traditional Shanghainese ingredients, including dates stuffed with glutinous rice paste, lily buds, and chopped celery. The diverse elements played off each other in a mouth-watering way: the crisp complemented the tender, the savory emphasized the sweet; the humble comfort of glutinous rice contrasted with the rich liver. Jereme's artful presentations also impressed me. Instead of simply molding his drunken chicken in a round bowl and inverting it onto a plate, as most restaurants did, Jereme boned and sliced the chicken, layered it over a mound of shaved ice spiked with rice wine, and served the pale slivers of poultry in a martini glass.

Jereme bounded in and out of the kitchen to check on the progress of my meal, and I was impressed by a chef who was so earnest about his food. After that first visit, I kept tabs on Jereme's progress. Patricia Wells called him a "genius" in the *International Herald Tribune. Saveur* ran a full-page picture of him. Matt Lauer visited the restaurant and ate Jereme's dumplings on *Today.* As I grew more adept at cooking, I began to dream that one day I might have a chance to work in Whampoa's kitchen.

I got the courage to approach Jereme after graduating from the cooking school and apprenticing at the noodle shop, which I hoped showed that I had endured enough *ku,* bitterness, to qualify me for an internship at the venerated Whampoa Club. With the help of mutual friends, I set up a meeting with him in one of his private dining rooms to discuss the idea. He had changed in the two and a half years since I'd last seen him. His youthful demeanor had been replaced with an intimidating sternness. His chef's jacket was embroidered with a new title: "Founding Chef." He had relinquished the day-to-day running of the kitchen to his sous-chef and was spending most of his time expanding the business, he said. His perspective had shifted.

"If a glass broke in the dining room [back when] I was cooking in the kitchen, I would think, how much did that impact the diner's experience? Now I have to ask, how much did the glass cost?" His years in Singapore had given his English a vaguely British lilt.

Sounding more like a CEO than a chef, he told me he'd recently started a company called Jereme Leung Concepts, which he hoped would make him a global brand. Restaurants would be the main business, but his company was working on other projects, including a line of tea, tableware, and wine.

So, he said finally, he was willing to let me intern, with one caveat: he insisted on approving anything I wrote about my experience before it was published. What if I saw a worm in the vegetables and decided to write about it? That would not be tolerated. He had to think about his image, he said. Not just for his sake, but for his investors'.

I agonized over the decision for days—dozens of restaurateurs all over China had already rejected me, after all, and where else would I find a place as highly respected as the Whampoa Club? Nevertheless, I told him as gently as I could that I would not take on an internship with strings attached. Oddly enough, he immediately acquiesced, as if he'd never made such a demand. I gathered that, in the end, having a journalist document his abilities was an opportunity he didn't want to pass up, particularly since he needed the media exposure in his quest to become a celebrity chef. It helped, too, that our mutual friends convinced the cagey chef that I wasn't a threat to his growing empire. A few days after our meeting, I was working in Jereme's kitchen with his blessing. "We have no secrets," he said with a poker face.

DRUNKEN CHICKEN (*ZUI JI*)

- 4 cups water
- 1 teaspoon powdered chicken bouillon
- ¼ teaspoon salt
- 1 teaspoon sugar
- 2 bay leaves
- 2 cinnamon sticks
- 3 star anise
- 1 tablespoon *baijiu* (grain liquor; popular brands include Maotai and Wuliangye; vodka may be substituted)
- 1 tablespoon *huadiao*, a high-quality yellow rice wine
- 3 thumb-sized chunks of ginger
- 3 scallions, knotted together
- 1 whole young chicken

In a small saucepan, combine the water, bouillon, salt, sugar, bay leaves, cinnamon, and star anise. Bring to a boil, then let cool completely. Add the *baijiu* and rice wine. Put the sauce in a large bowl.

Fill a large pot with water, add the scallions and ginger, and bring to a boil. Add the chicken. Cook for 30 minutes, or until chicken is done. Plunge the chicken into ice water to prevent the skin from falling off.

Bone the chicken and cut it into bite-sized pieces. Add the chicken pieces to the sauce and let them marinate for about 10 hours, or overnight. Before serving, drain the chicken and serve it in a bowl.

Jereme Leung's variation: Make an extra batch of sauce and freeze it in ice cube trays. Just before serving the chicken, shave the iced sauce and layer the chicken pieces over it in a martini glass. Serve immediately.

Three days a week, I reported to the restaurant at ten o'clock in the morning and worked through the lunch shift, which lasted until three. I entered the building through a back alley, along with the chefs, suppliers, and service staff. I slipped on my chef's jacket and pants in the locker room, pulled my hair back into a ponytail, and donned a paper hat alongside hostesses in elegant *qipao,* who preened and blotted their lipstick in front of a mirror. An apron and tennis shoes completed my outfit. I would have preferred a pair of cooking clogs like the kind Jereme wore, but they weren't for sale in China. Most of the Whampoa chefs couldn't have afforded them anyway—they would have cost more than a quarter of their average monthly salary, which ranged between $200 and $500. So most of the chefs wore black cloth shoes, which seemed unwise in a place where a cleaver could come crashing to the ground at any time.

In the kitchen, I shelled beans and wrapped strands of noodles around single servings of shrimp, to be presented as an amuse-bouche. In my more glorious moments, I wrapped *xiao long bao*. But mostly I trailed the chefs, observing their every movement. I was exhausted by the end of my shift, which was only the halfway point in the day for the rest of the chefs; after a two-hour break, they'd return for the evening shift.

I gathered that Jereme had put his recently promoted executive chef in charge of keeping me busy and distracting me with samples of food so I didn't go poking in too many corners. The top chef was a Malaysian of Chinese heritage in his early thirties. His name was Hew Choong Yew, but in the kitchen, the Chinese staff called him by his Mandarin name, Brother Yao. He had worked with Jereme for several years before coming with him to set up the Whampoa Club.

Brother Yao's tall, lanky frame gave him a good view of all the goings-on, and sometimes he'd leap suddenly across the kitchen. Some remarked that his moods mirrored Jereme's, which "changed like the weather." One moment, Brother Yao could be as goofy and amiable as the Swedish chef on *The Muppets*, and the next moment, if Jereme expressed disapproval of something, he could turn as icy as an off-the-air Martha Stewart. His usual post was in the middle of the kitchen, in the wok department, where the main courses were stir-fried, seared, and stewed.

Brother Yao introduced me to the wok stations as if he were a car dealer showing off the latest models. "We got these from Guangzhou," he said as he strolled past one station, his kitchen towel swinging from his waist. "They were made for export to Hong Kong. The quality of most of the local ones is pretty bad. They sound like this." He picked up a tin bowl and rapped on it with his fist.

The wok chefs sweated heavily through their whites. The

giant woks they lifted and shook weighed close to ten pounds each, and the fires beneath them, produced by six gas burners per station, were intense enough to sear food in a matter of seconds, sealing in the flavor. The size of the fire could be adjusted with a foot pedal; pushing down on the pedal raised the flames and kicked the overhead fan into overdrive, creating a roar that sounded like a truck's engine. When the chefs weren't cooking, the fire could be lowered to a whisper, but it never went out until the shift was over. Each burner was elevated, and the base was essentially a large, shallow sink where chefs dumped leftover bits of food and excess oil, minimizing the time spent cleaning the woks. Usually, all they got between dishes was a splash of water from the tap attached to each station and a swirl with a bamboo brush.

Everything else was cleaned assiduously. Dishwashers worked on one side of the kitchen, and the cooks who washed vegetables, meat, and seafood were stationed on the other. In between, a row of stainless steel counters ran the length of the kitchen, each a little island devoted to a particular preparation method that contributed to an elaborate Chinese meal.

To the right of the woks was what Jereme called the cutter department, where chefs scrutinized and boned the meats and fish and sliced and diced the vegetables. Much of the seafood arrived alive. Fish tanks occupied one wall, bubbling with brightly colored sea life. Orange hairy crabs — small crustaceans with furry patches on each claw that resembled pompons — attempted to scale the slippery glass walls, seemingly aware that their lives were in imminent danger. After the crabs were gone, a couple of rosy Australian lobsters took their place. They held each other's claws as if they were ballroom dancing, unaware that their courtship would soon be cut short.

Sitting next to the tanks was a Styrofoam box filled with croaking bullfrogs.

To the left of the wok chefs worked the appetizer chefs. Though their department was small, consisting of two counters and a lone wok station, it churned out some of the restaurant's best-known dishes: the memorable drunken chicken and other favorites, like Shanghainese smoked fish and candied lotus roots with glutinous rice. The appetizer department abutted the steamer area, where chefs busily fed a constant stream of *xiao long bao* and other local specialties into a steamer that resembled a tall oven with numerous racks.

Next to the appetizer and steamer sections were separate dim sum and dessert alcoves. It was rare for a Chinese kitchen to have a dessert department; traditionally, meals ended with slices of fruit rather than plates of sculpted sweets like curried chocolate ice cream or ginger crème brûlée. Occasionally even the sophisticated Whampoa Club stumbled here: its chocolate noodles—limp, doughy strands tasting faintly of cocoa—were quickly scrapped from the menu.

Jereme's exacting standards were everywhere in evidence. Prep chefs carefully measured ingredients on a scale. Copious amounts of plastic wrap, enough to mummify everything in sight, were used to cover food that was not served immediately. A lavish supply of hand soap in plastic dispensers was fastened to the wall above each sink. After the cooking school and the restaurants in which I had worked before, where chefs and waitresses handled food with dirty fingers, in flagrant violation of the most basic rule of hygiene, the soap was a novelty. It relieved me of the almost criminal feeling of setting to work without lathering up.

The chefs labored in first-world conditions. Compared to

the standard Chinese kitchen temperature of around 100 degrees, the Whampoa maintained a cool 77 degrees, thanks to a thermostat. Typical Chinese chefs worked seven days a week, with a day or two off each month if they were lucky; the Whampoa chefs had two days off each week. At many Chinese restaurants, bosses served their workers scraps they couldn't get away with selling to their customers; the Whampoa Club had a clean cafeteria where the staff ate tasty, balanced meals.

Jereme sought out the best ingredients, wherever they happened to be. The Whampoa chefs cooked with *huadiao*, a high-quality rice wine from nearby Shaoxing. The standard 40-cent bottle of cooking wine available at every supermarket, a staple of my Beijing experience, didn't have the body or richness of *huadiao*. The restaurant served Thai jasmine rice, which was more fragrant and had a springy, tender texture that Chinese rice rarely achieved. Jereme imported his potato starch from the United States, because Chinese potatoes weren't starchy enough. They were too crisp, "like crackers," he said. Brother Yao instructed me to use imported Sunkist lemons, not Chinese lemons, if I wanted to duplicate the recipes at home. Jereme lamented the lack of good beef in China, and the excessive amount of tenderizer Chinese chefs relied upon, and the next time I ate beef in a restaurant, I noticed its artificial, almost pulverized texture. One traditional ingredient that I did find in the kitchen was MSG. I spied Brother Yao handling a container of the taste enhancer one morning, and he didn't look bashful about it. "Gotta make the customers happy," he explained with a wink.

Aside from the inescapable kitchen hazards—the errant knife or the squeaky-clean floors, mopped so obsessively that they were often slippery—I felt that I had been magically transported into the developed world, a China of the future, where rules were followed, quality was valued, and the life of a chef felt humane. The attention to cleanliness and quality, of course, came at a price. A meal at Whampoa cost an average of $50 per person, in a city where a traditional sit-down dinner was less than $10 a head.

There was something more that set the restaurant apart, however. Slave driver that he could be as a boss, Jereme was also a source of inspiration for his chefs. They called him Laoda, a term of respect given to the eldest sibling of a family, and they seemed particularly impressed that he had risen to such culinary heights from a humble background. "I hear he's bought a house!" one of his chefs whispered to me one morning.

So how did a boy who grew up outside mainland China, with little knowledge of Shanghainese food, end up becoming a pioneer in the city's booming restaurant scene? Cooking ran in his blood, for one thing. Jereme's father had been a cook on a cargo ship. His parents divorced when he was young, and he spent much of his childhood with his maternal grandparents, who were restaurateurs. They had started by selling wontons from a cart in the streets of Singapore. Eventually they traded in their cart for a *da pai dang,* a stall in a traditional outdoor food market. As business improved, they opened three outlets around Singapore and then expanded to Hong Kong, where Jereme had been born and raised.

The restaurants were Jereme's playgrounds, and his grandparents doted on him. They would chop off the juicy drumstick of a Cantonese roast duck and hand it to him as a treat. He remembered being surrounded by food: rice porridge, fish steamed in a simple soy broth. The alcohol flowed. His grand-parents allowed young Jereme to take shots of *baijiu*, brought out as an offering during rituals of ancestor worship. "My family was not well educated, not very refined," Jereme remem-bered. "I would walk into a wall and wonder, Why can't I cross this?" He let out a string of his trademark giggles, bubbling with nervous energy. "I'm a trained alcoholic."

Jereme spoke more vaguely of his adolescent years, when he lived with his mother. Family tensions ran high after his mother remarried, though he didn't elaborate. "Sometimes there are certain situations in a person's life . . ." He trailed off uncomfortably. "So I left the family."

Jereme moved out at thirteen and dropped out of school, an institution he recalled with bitterness. "I hate school," he mumbled sullenly. "I was not good in a traditional environ-ment. I remember, on a mathematics test, I once scored seven out of a hundred. The teacher probably gave me seven points because he was too embarrassed to give me a zero." School "didn't work with me. I don't learn that way."

He went to work as a dim sum apprentice at a restaurant owned by his grandparents on a crowded street in the British colony. Once he had learned the ropes, he hopped around to various establishments, going to whatever restaurant would give him work and a bed. The restaurants were Cantonese, considered the most refined of China's cuisines. All over Asia, Cantonese was the type of food that Chinese chose to eat on special occasions, celebrated for its seafood and delicacies such as shark's fin soup, braised abalone, and steamed scallops. In Hong Kong, Cantonese staples included an array of barbecued meats (goose, duck, pork), pan-fried noodles, and dim sum items like egg tarts and shrimp dumplings.

After five years, Jereme had become adept at the three pillars of Cantonese cooking: dim sum, barbecue roasting, and wok frying. "I could cook a proper banquet then, but I was not experimenting yet. I was a very traditional Cantonese chef. If I had experimented, I would have been fired," he said.

Jereme's kitchen career was interrupted at eighteen, when he went to join the Singaporean army, a service that would grant him a passport to the island city-state. When I asked him why he had wanted a Singaporean passport, he answered evasively. "My mother wanted me to have it." I suspected he was being careful not to offend his Chinese customers or the Chinese Communist Party. At the time, many Hong Kong residents were seeking citizenship elsewhere—apprehensive, somewhat justifiably, of the consequences of authoritarian China's imminent reclamation of the British territory in 1997. At that point, Jereme had no way of knowing that, more than a decade later, he would become a restaurateur in mainland China.

Quite unexpectedly, Jereme's two years in the military turned out to be critical to his later success. He was posted to a unit that contained some of the army's most educated recruits: Singaporeans who had received scholarships to go overseas for master's degrees and deferred conscription until they graduated. "We were all there because the government didn't know what to do with us," he joked. The recruits, representing Singapore's diverse ethnic groups, including Indian, Malay, and Chinese, spoke English to each other. Fraternizing with them, Jereme learned to speak the language fluently. He also studied for and passed the British version of a high school equivalency exam.

Jereme speculated that his superiors, sick of the usual mess hall meals, had placed him in the special unit because of his cooking skills. "The army chefs made big pots of shit. So when there was a banquet for the generals, the Ministry of Defense would ring me up and I would do all the cooking."

One of Jereme's bunkmates had just returned from studies in France and introduced Jereme to wine. The bunkmate also had an appreciation for food, and together they dined everywhere, from Malaysian street stalls to the island's finest European restaurants. One night they went to the restaurant in the Meridien Hotel, where he tried foie gras terrine for the first time. "Didn't like it. It's cold. It's liver. Chinese are used to hot food. It was a piece of cold, mashed-up meat," Jereme said. The foie gras chef, who had received three Michelin stars and was visiting from Europe, came by the table and asked Jereme how he liked the dish.

As he recalled the incident, Jereme shook his head and squinted. "I said, 'It's no good!'" He let out a roaring giggle, then turned serious. "I had read about proper etiquette in books, but it was entirely different trying it out in a restaurant." He said that his humble roots made him more patient with his staff and their occasional ignorance about fine dining. "It's not their fault. It later becomes a burden to admit that you don't know about those things. I was lucky to have that exposure then. Those two years shaped me."

I started my internship in the Whampoa Club's dim sum room, where I learned the secret of making *xiao long bao*. The dim sum room made other savory snacks, like deep-fried dumplings with a fluffy filling of shredded white Chinese radish, which was savory-sweet, with none of the bitterness of its American counterpart. The pan-fried pork dumplings were crisp and juicy. But the dumpling that held my attention most was still the *xiao long bao*. They were a completely different breed from the crescent-shaped dumplings I had learned from Chairman Wang. *Xiao long bao* were round, with folds forming a spiral on their tops. They were steamed rather than boiled. Though the wrapper was thin, it was strong enough to hold the pork-based filling and the wonderful buttery broth that spurted out when you bit into them. "You'll never want to eat them again after I tell you what's in them," Brother Yao said before he divulged the secret: the soup inside the delicate skins was made from jellied pork skin.

I watched a chef make pork-skin jelly one morning. The chef, a polite Cantonese man in his twenties who tolerated my constant questions, began by hacking large sheets of pork skin, half an inch thick, into six-inch-square pieces. He put them in a tall pot of boiling water, about a dozen at a time, adding *huadiao* (cooking wine) and white vinegar. After ten minutes, he removed the skin from the pot, transferring it to a shallow tin pan.

When the skin had cooled, he lit a cooking torch—the kind used in restaurant kitchens to caramelize the top of a crème brûlée—and ran the bluish-orange flame across the skin to scorch off any remaining bristles. Then he chopped the skin into smaller pieces and sliced away the fat underneath. This took a couple of hours.

I couldn't help noticing that the skin was stained with blue ink, presumably a stamp from the slaughterhouse.

"The ink doesn't come off?" I asked.

"No," he said. "*Mei wenti*—it's no problem." Having cooked the skin beyond 212 degrees, it was sterilized, he assured me. But I was unconvinced. Yet another reason not to eat *xiao long bao*.

The chef simmered the defatted skin with chicken stock—

one part pork skin to eight parts stock—for four hours, until it dissolved into a gelatinous, gooey mess. The chef let it cool, then pulsed it in a blender and pressed it through cheesecloth. He added a sprinkling of chicken bouillon and put the mixture in the refrigerator, where it would solidify like Jell-O overnight.

The next day, the chef spooned several dollops of the jelly into a small bowl of minced pork and whisked it until the mixture was smooth, like cake batter. This would be the filling, and when the dumplings were steamed, the jellied pork-skin broth would liquefy and seep into the delicate wrappers. The dough for the wrappers was a little different from what I was used to as well. In addition to flour and water, a large spoonful of pork fat was folded in, which made the skin more pliable and tender. Now I knew why *xiao long bao* tasted so good—every element was imbued with the fatty deliciousness of pork.

Brother Yao poked his head into the alcove to demonstrate the wrapping technique. Holding a skin in his palm, he spooned a ball of the minced pork filling onto the wrapper, then used the thumb and forefinger of his other hand to bring the edges of the skin together, pinching the dough to seal in the filling while spinning the dumpling in his palm so that the creases wound around each other, forming the distinctive spirals.

"It's like turning a screw," Yao said, his palm revealing a perfectly wrapped sample before he disappeared back into the main kitchen.

Then it was showtime, my first chance to wrap the legendary *xiao long bao*. Or so I thought, until the chef handed me a ball of the gummy, wax-like material chefs used to anchor serving glasses on plates. "What is that?" I asked him. He indicated that I was to use it to practice, rather than waste real meat on my shoddy workmanship.

I fumbled, making uneven folds. When I finished, I picked up another skin and tried again, adding a second layer of skin around the fake filling. When I had covered the dumpling in multiple layers of skin, I started with a new ball of gummy wax. The mock dumplings multiplied, each encased in several shells like Russian dolls.

Toward the end of the shift, the dim sum chef presented me with a steamer basket. He lifted the lid to reveal a trio of *xiao long bao*, their translucent skins glistening with steam. I didn't hesitate: I snapped one up with my chopsticks. I nibbled the skin to pierce it, then poured the broth into a porcelain spoon and savored it. Brother Yao was wrong. My desire for *xiao long bao* hadn't been dampened at all.

SHANGHAI SOUP DUMPLINGS
(*XIAO LONG BAO*)
Makes about 3 dozen dumplings

PORK-SKIN JELLY
- 1 pound pork skin
- 1 tablespoon vegetable oil
- 3 slices of ginger
- 2 scallions, chopped
- 1 quart water
- ¾ cup Shaoxing rice wine or sherry
- ¾ cup white vinegar
- 2 cups chicken stock
- ½ teaspoon powdered chicken bouillon

DUMPLING FILLING
- 1 pound pork belly, minced
- 1 large egg
- 1 teaspoon minced ginger
- 1 tablespoon sesame oil

¼   teaspoon salt
½   teaspoon sugar
½   teaspoon powdered chicken bouillon
¼   teaspoon ground white pepper
1   teaspoon light soy sauce
2   teaspoons dark soy sauce
1   batch pork-skin jelly (see recipe)

DUMPLING WRAPPERS

3   cups all-purpose flour
1   cup high-gluten flour
1½ cups water
⅓   cup lard or vegetable shortening
    Black vinegar and shredded ginger, for garnish

To make the pork-skin jelly (must be made one day in advance): Blanch the pork skin in hot water for 2 to 3 minutes, drain, and allow to cool. Meanwhile, place the oil in a wok over high heat. When the oil is hot, add the ginger and scallions and stir for 2 to 3 minutes. Add the water, rice wine, and white vinegar and bring to a boil. While waiting for the mixture to boil, burn any remaining bristles off the pork skin with a cooking torch, or pluck out the hair with tweezers. Slice the cooled pork skin into 4-inch squares and add them to the wok. When the mixture begins to boil, reduce the heat to medium, cook for 15 minutes, and discard the liquid. In a separate pot, heat the chicken stock. While waiting for the stock to boil, trim any excess fat from the inside of the pork skin. Add the skin to the chicken stock and simmer for 4 hours. The skin should dissolve in the chicken stock. When it has cooled, put the mixture in a blender, and blend until smooth. Strain the liquid through a sieve into a bowl. Stir in the bouillon and place the liquid in the refrigerator overnight, or until it hardens into a jelly.

To make the filling: Combine the pork belly, egg, ginger, sesame oil, salt, sugar, bouillon, pepper, and soy sauces in a mixing bowl, stirring in one direction for 50 strokes. Just before wrapping the dumplings, take the pork-skin jelly out of the refrigerator and mix it in.

To make the dumpling wrappers: In a large bowl, combine the flours. In a separate bowl, combine 1 cup of the flour mixture and ½ cup of the water. Using your hands, mix the flour and water together until it is uniformly blended. Add the rest of the flour and the water, a little at a time, mixing until a dough forms. Add the lard and knead the dough until it is evenly dispersed. Divide the dough into three equal sections, and roll each section into ropes. Divide each rope into about a dozen pieces. Flatten each piece with your palm, sprinkle flour over the dough, and roll it out with a rolling pin into a thin, circular shape slightly larger than your palm. Sprinkle the circles with flour.

To wrap the dumplings: Place a dumpling wrapper in your palm and spoon a dollop of the pork mixture into the center. Using your thumb and index finger, pinch together the edges of the wrapper at the top of the dumpling, and rotate the dumpling in a circle to create a spiral.

Steam the dumplings over high heat for 7 to 8 minutes. Serve immediately with black vinegar and shredded ginger.

12

ON ONE OF MY DAYS OFF from Whampoa, I phoned my friend Jiang Liyang, a fellow food critic, and invited him to lunch at Crystal Jade, my favorite dim sum place in Shanghai. It was just before noon, and the stylish restaurant, decorated with wooden lattice screens and tall windows, was already filling up. My sixty-two-year-old friend, wearing his trademark suspenders, shuffled toward the table, greeting me with a smile and a chuckle that shook his belly. He had large, squarish black eyeglasses, a bulbous nose, and a wide mouth that I knew from experience could shovel in epic quantities of food.

Jiang had scarcely sat down and unfurled the paper fan he always carried when the manager came by to say hello. Restaurant proprietors often interrupted his meals, if not in person, then via Jiang's cell phone, which seemed to ring every few minutes with the promise of a free meal. He showed off his new Nokia, which featured a graffiti pen and a touchpad. "This only cost me a thousand *yuan!* You should get one," he said, eyeing my outdated model sitting on the table.

I noticed that Jiang's suspenders were hanging a bit slack over his shoulders. He had once explained that he wore suspenders because they allowed him to keep wearing the same pants, whatever his weight. At five foot six, he had weighed 207 pounds at his peak, but now his weight fluctuated between 160 and 190. At the moment he was a svelte 170. He had made an effort to lose weight ever since he had developed type 2 diabetes a few years before. Like many urban Chinese who had prospered in the past two decades (his monthly income of $1,200 put him solidly in China's upper middle class), he had succumbed to the disorder after an adolescence marked by malnutrition gave way to an adulthood full of fat, sugar, and carbohydrates. But his gluttony persisted, albeit in a sugar-free mode.

I had befriended Jiang early in my food-writing career, when I was living in Shanghai. Like me, Jiang was freelancing for a number of publications, including the *Shanghai Tatler* and the Chinese edition of *Vogue*. We had bonded over our mutual interests, and I used the honorific "Teacher" as a sign of respect for his age and knowledge of food.

In Teacher Jiang's opinion, ham tasted better when smoked with dog meat ("I know foreigners don't like hearing that"). When he ordered pigeon, he started with the left leg, which he believed was the tastiest part of the bird, because (according to Jiang) pigeons put more weight on their left limb as they strutted, making the meat gamier and more flavorful. He didn't understand why veal was prized in the West: "If the animal doesn't move, how can the meat taste good?"

At Crystal Jade, he urged me to try the eyeball of a "fat-head fish," the Chinese name for a variegated carp. The eyeball was part of a dish called "Hunanese chopped chili fish head." The head, as large as a rib eye steak, was served on a platter garnished with red peppers and scallions. It was butterflied, making it easier for the waiter to scoop up the head and plop half of it on each of our plates.

Fish head was universally loved by the Chinese, and having suffered through many Chinese banquets, I had acquired a taste for the delicacy. Once I stopped thinking about what I was eating, I enjoyed the range of soft textures, especially the tender cheeks. The head had a sponge-like capacity to absorb whatever flavors it had been simmered with. But I hadn't yet overcome my aversion to the eyeball.

"It's the best part!" Teacher Jiang said with his usual punchiness. He had demolished his portion and was sitting back fanning himself, waiting for me to finish.

"What did you think of the fish?" I asked, stalling.

"It's very tender. The flavors are light and balanced."

"Anything about it that you didn't like?"

A waitress came by to remove his plate. My plate, with the fish head, remained.

"It's not very authentic," he said. "It's not as spicy as most Hunanese chopped chili fish heads."

But if it tasted good, what did it matter whether it was authentic or not?

"They call it 'chopped chili fish head' on the menu. So if they call it that, it should be the real thing. Changing it, refining it, is fine. But then they shouldn't call it the name of a Hunan dish." He paused, searching for an analogy I would understand. "You can't call something a Clinton then give them a Bush!" Like many Chinese, Jiang had a fondness for Bill Clinton that George W. Bush didn't inspire.

Finally I got up the nerve to bring the eyeball to my lips. It stared blankly back at me. I quickly shoved it in my mouth and chewed, taking in a gelatinous texture similar to that of a supreme cut of *toro,* raw tuna belly. I swallowed and spit out a hard, pearly bit that must have been the socket.

"See?" Jiang bellowed and laughed. "It was good. You'll be fighting for it next time!"

Jiang, like many Shanghainese, had a love of anything new. He embraced technology with gusto and had started a *bu ke*—a blog—about food, which he wanted to turn into a profitable venture. He posted items about the latest counterfeit foods he had found in markets and restaurants. Because of lax regulations, Chinese consumers sometimes came across products that were advertised as something they weren't—a cheaper fish labeled as shark's fin, baby food lacking the proper nutrients.

Jiang also enjoyed trying new foods, unlike many Chinese I had met. I found it ironic that some Chinese had no qualms about diving into jellyfish, stinky tofu, and donkey penis but balked at spaghetti and meatballs. Many Chinese tourists, when traveling overseas, insisted on eating Chinese every meal, even in culinary paradises like France and Italy. But Shanghainese were more open-minded. The city, with its colonial past, had always had more contact with the outside world. That openness, combined with its newfound prosperity, helped chefs like Jereme thrive in Shanghai. Jiang had dined at the Whampoa Club a number of times and was impressed by Jereme—"He can really cook like a Shanghainese!"—although he pointed out that the restaurant was too pricey for the average citizen to enjoy.

The next time, we met for lunch at a Continental restaurant. The walls were painted in muted gray and blue hues and illuminated by track lighting, and there was an un-Chinese absence of noise. The food critic had brought along a friend named Zhuang Jian, who wore a button-down shirt and slacks and had the bland demeanor of an accountant. With single-portion dishes and wide tables, the restaurant made it challenging to eat family style, but we pushed the plates back and forth so we could share everything, the way Jiang and I always did when we ate Chinese food.

Though he was more open-minded than most Chinese, eating in the Western restaurant was still a novel experience. He prodded and poked the thin slices of beef carpaccio and smoked salmon with his fork, holding them in the air for examination as if they were laboratory specimens. Yet he dove eagerly into every new dish, while Zhuang took tentative nibbles and set his fork down as he chewed.

Zhuang passed me his business card, which said he was deputy secretary-general of the Shanghai Restaurant Association. Chinese associations were a mystery to me. Often their sole purpose seemed to be to give people another title to cram on their business cards. Nearly everyone had at least one, and people insisted on exchanging them with everyone they met. The Shanghai Restaurant Association's purpose was to "share information and hold meetings," Zhuang explained. His card didn't impress — he had only one other title: deputy general of the organizing committee for the 2006 Shanghai International Restaurants Exposition.

Zhuang liked numbers. Shanghai was home to 40,000 restaurants that collectively served the cuisines of more than 40 nations, including Brazil, Turkey, and Nepal. The largest restaurant in Shanghai had five floors of dining space totaling 100,000 square feet, which included 108 private dining rooms, and generated $23,000 in revenue per day.

He noted that all kinds of businesses ran restaurants on the side—not just hotels, but bathhouses, spas, and saunas. His association counted takeout places as restaurants. Some takeout operations catered to elementary and middle schools, which often didn't have cafeterias of their own. The largest of these operations sent out 100,000 lunches per day, and the association pressed them to make their meals not only nutritious but tasty. "If they don't like the food, kids will go and buy lamb skewers on the street," Zhuang explained.

I asked how much a school meal cost.

Zhuang's reply was a little long-winded. Shanghai, like most of urban China, had the "four-zero five-zero problem," he said. He was referring to people in their forties and fifties, who were born around the time of the Great Leap Forward and didn't receive a proper education because of the Cultural Revolution. (Chairman Wang was one of the older members of this group.) After working in factories for a couple of decades, many of these urban middle-aged workers had been laid off, as factories in city centers like Shanghai and Beijing relocated to southern and rural parts of China. "And they have kids, but they can't afford to raise them. They still have to feed their children *and* their parents. So we've set the price at four and a half *yuan* per meal."

Zhuang mentioned that many Shanghai restaurants served China's "five delicacies"—sea cucumber, shark's fin, abalone, fish maw, and bird's nest. When I asked him how many, he paused, looking befuddled. That was one number he didn't know. To help him save face, I quickly asked if shark's fin was still popular. I had never understood the appeal of the tasteless, translucent fibers.

"Eating shark's fin increases your internal heat too much.

It's not good for the environment. Yao Ming doesn't eat it," he said, referring to recent public service announcements against shark's fin that featured the Houston Rockets basketball star. The Chinese, who had been conditioned by years of dogmatic government propaganda slogans, seemed to have found it refreshing to be lobbied by a celebrity.

Zhuang added, "There is a lot of fake shark's fin on the market."

Jiang looked up from his food as if breaking from a trance. "It's just like mooncakes," he said.

The midautumn Moon Festival was a few weeks away. The holiday fell on the full moon of the lunar calendar's eighth month. The evening was celebrated by eating dense round cakes with sweet or savory fillings. The way Jiang described them, it sounded as though mooncakes had become the Chinese version of leaden Christmas fruitcakes. "Nobody buys mooncakes for the sake of eating them. Some are made from low-quality ingredients, or recycled from the year before. We just send them to people. There is too much sugar in them."

"I like mooncakes," I said. Or at least I thought I did—I had fond memories of the snack from my childhood.

Good, Jiang said. He opened his briefcase and handed me a pair of vouchers that he had received as a gift. I could redeem them at a local bakery for two boxes of the pastries.

China had adopted new regulations on mooncakes, Zhuang added. "They can't be as ostentatious as they used to be." By law, the total value of the box itself couldn't exceed the value of the mooncakes it contained.

Why would the government want to micromanage mooncake packaging? I asked.

In recent years, Zhuang explained, mooncakes had become

a medium for transporting bribes. Boxes made of 18-karat gold had begun to appear. Sometimes wads of cash were hidden inside the cakes. "Or keys to a new apartment, or a BMW. Mooncakes have become a tool of corruption," he said.

I had heard a legend that seven hundred years ago, during the Yuan Dynasty, a Chinese general had toppled the ruling emperor by sending his soldiers messages embedded in mooncakes. Somehow this modern use of mooncakes didn't have the same heroic ring.

Zhuang's mention of corruption riled Teacher Jiang. Why, just yesterday, he bellowed, a top-level Communist Party official from Shanghai had been sacked for stealing millions of *yuan* in government funds. It was the first time such a high official would face a trial and prison time. "He deserved to be sacked," Jiang said in his usual outspoken manner. "He took our pension funds for his own use. He took our money!"

Zhuang shifted uncomfortably in his seat and returned to his recitation of facts and figures. In 2005, the average restaurant in Shanghai had revenues of around three million *yuan*. In 2006, the figure was expected to increase to four million *yuan*. "Our country doesn't have the best agricultural technology in the world," Zhuang said. "We don't have the most industrialized economy. But we have some of the world's best cuisine."

Toward the end of the meal I could tell that Teacher Jiang was still hungry, and I guessed that it was because it hadn't ended with something starchy. I asked the waitress to bring some pasta — the closest thing I could think of to Chinese noodles. She reappeared with a plate of fettuccine in light cream sauce with duck confit. Jiang happily devoured the dish.

· · ·

Like most food writers in China, Teacher Jiang accepted free meals when he reviewed restaurants. Restaurants sometimes paid him to write advertorials too. Proprietors asked for his advice not only on how to improve their food but also on their marketing strategies. In China, business interests, along with directives from the Communist Party, often came before journalistic objectivity. Jiang maintained that the free meals and advertorials didn't affect what he wrote. "If I don't like the food, then I just won't write the review," he said. "I've turned down money from restaurants in some cases." But he looked uncomfortable as I continued to probe, so I decided to drop the subject.

Maybe I suspended my judgment about Teacher Jiang's conduct because, after spending time in China, I realized that I couldn't hold people there to the same ethical standards that applied in the United States. It didn't surprise me, after learning how Chinese journalists operated in Beijing. But it was also because I just plain liked Jiang. He was outspoken, and I admired his unadulterated love for food and writing.

His passion for food began at an early age. Jiang was born in 1944 and was the oldest of six children. His mother was a vegetable merchant, and he remembered her cooking fondly. "She made bean sprouts, using everything down to the root. What most people would throw away nowadays, she made into a perfectly good dish." She used *congbai,* a fragrant, scallion-like vegetable, in her red-braised pork. The key to her delicious chicken soup—"just pure chicken"—was her skillful control of the fire.

His father built water heaters for a living, and the family led an average, middle-class existence in a *shikumen* row house in downtown Shanghai through most of the 1950s, when food was relatively plentiful. But by the end of the decade the meat they were used to eating began to disappear, and Shanghai lapsed into poverty, along with the rest of the nation, under Mao's Great Leap Forward. Jiang remembered that his family often could afford to burn only one 25-watt bulb at night. He sometimes stayed up late writing in the semidarkness anyway. In high school he won an essay competition sponsored by the *Xinmin Evening News,* one of Shanghai's most influential newspapers. His essay was published, and he was awarded 40 cents, which he used to buy a new T-shirt, to replace his old, tattered ones.

During his teenage years the family mostly ate rice and noodles. "We had so little nutrition that my brain felt a little foggy." His family of nine—six siblings, parents, and one grandparent—received two eggs per day. "My parents made me eat all the eggs. They figured that if I ate well, I could test well on the college entrance exam. If I went to university, there would be more room in the house. There would also be one less mouth to feed."

The plan succeeded. In 1962, Jiang was awarded one of thirty places in the class entering the journalism program at one of China's top schools, Fudan University, in the northern suburbs of Shanghai. He recalled the food, which was a step up from what his impoverished family could provide. "There were even dishes with meat *and* tofu. We had fatty pork and rice porridge for breakfast. We sometimes ate steamed fish."

Jiang spoke less comfortably about the philosophical clashes he had with his professors and classmates, just as the Cultural Revolution was getting under way. He wrote his thesis on his favorite book, the eighteenth-century novel *The Dream of the Red Chamber.* His fellow students harassed him, demanding to know why he was reading old classics rather than Mao's philosophy.

"I told them, 'Mao's an artist. He shouldn't be involved in politics.'"

Jiang's tastes and his blunt opinions on the Great Helmsman were grounds for punishment in the hysterical, Orwellian era that China had entered. When his classmates graduated, they became government mouthpieces at state-run newspapers. The school's officials held him back at the university to perform a year of manual labor. Then he was sent to Anhui, a rural province that borders Shanghai, to teach in a high school.

"I started to research food when I was in Anhui," he said. "I got a stove. I found a cookbook that described cooking techniques used by an emperor's chef from a thousand years ago." In the lush Yangtze River Delta, where Jiang lived, the food was plentiful. "I could eat fish, shrimp, and crab. I made my own sausages." He realized that by researching food, he could satisfy his hunger, both physical and intellectual, and "no one would bother me. This was a safe thing to study. I was very happy."

After Chairman Mao died in 1976 and the Cultural Revolution ended, the authorities decided that Jiang had been sufficiently "reeducated." He returned to Shanghai, earned a master's degree in journalism from Fudan, and was put on the business beat at *Wen Hui Bao,* a large Shanghai daily. Along with the rest of Chinese society, newspapers were undergoing reforms and had adopted new techniques. "We used the inverted-pyramid method of news writing," he told me, then launched into a full explanation of it. I was too polite to mention that I'd had the lesson in high school journalism class.

Jiang wrote about China's skyrocketing first-time purchases of televisions and refrigerators. He also wrote about textiles and clothing. But he never really cared about anything besides food, so he steered his reporting in the direction of restaurants, agriculture, and food, convincing his editors that they were serious business topics. When he retired in 2004, he turned to food writing full-time.

By the time I met Jiang, the Cultural Revolution was long over. People could freely read *The Dream of the Red Chamber* once again, and the government was loosening up enough to tolerate his small exposés of fake foods. But writers could still be persecuted for criticizing government figures in print, on television, or in cyberspace, and as he wrote about the offerings of Shanghai's grandest restaurants, I wondered if the gourmand was aware that, on some level, he remained a writer in exile.

One morning, I went to redeem the mooncake vouchers Jiang had given me. It was just days before the Moon Festival, which according to legend originated sometime during the Han Dynasty (206 B.C. to A.D. 220), when Chinese elders decided that the moon that rose in the middle of the eighth lunar month was the roundest and brightest of the year. The holiday usually fell in the early autumn of the Gregorian calendar, and rural communities saw the coming of the festival as the start of the fall harvest. Many Chinese continued to be avid moon gazers and followed the lunar calendar. The holiday was celebrated by eating mooncakes and admiring the moon, preferably from a hilltop.

Though my family didn't go moon gazing when I was growing up in San Diego, we kept some lunar traditions. Sometime in my teenage years, my father, in a bid to reclaim his Chinese roots, announced that he was reverting back to the lunar calendar and would celebrate his birthday on its lunar date—which drove the rest of the family crazy, because we could no longer keep the date straight. Mooncakes were also important; my mother bought them at the Chinese grocery just before the festival each year. The cakes always came in a package of four, and were covered with a crust of flour, sugar, and lard set in ornate, ridged molds that featured designs of auspicious Chinese characters. Inside the dark golden crust was a dense filling that could be sweet or savory. I loved the cakes filled with lotus-seed paste—biting into one swathed my tongue in a sweet and smooth bath with a tea-like flavor. Each mooncake probably contained close to a thousand calories, and, well aware of this, my mother would cut each into eight wedges, which lasted us several days. When we were done eating one, she would take the tin box from the refrigerator, remove another cake, and cut it the same way. It would take us two weeks to finish the box.

On the way to the bakery, I noticed mooncakes everywhere. As I got off the elevator in the apartment building where I was staying with a friend, I stepped around an incoming stack of mooncake tins, with their trademark square shape and gold trim, piled so high that it obscured the man carrying them. Walking down Shanghai's chaotic streets, I saw crates of mooncakes crammed in restaurant windows. Even Starbucks was in the business, as I discovered when I stopped in one of Shanghai's many outposts of the chain. The barista asked if I wanted a cappuccino-flavored mooncake to go with my coffee.

A block away from the bakery, on a side street shaded by trees, a man lurked before a sign that read MOONCAKE PICKUP SITE. Gesturing to the vouchers in my hand, he asked, "How much?" He offered 34 *yuan,* or about $4, each. The face value of the voucher was double that.

"Sorry, they're not for sale," I said. As I continued on to the bakery, I was stopped by a scalper every few yards. The next two dealers jacked up the offer to 40 *yuan*. When I kept walking, insisting that I might want to redeem them myself, the fourth scalper laughed. "Okay, one last offer. How about eighty-five *yuan* for both?" He wore a white undershirt and sweatpants and carried a black clutch purse under his arm. He trailed me, in the way that persistent Chinese with an agenda usually did, leaving little space between us, until we were in front of the bakery.

"Really," I sighed. "I might want to eat them."

"Who eats them anymore?" he said.

"Aren't they any good?" I asked.

"Of course not. They don't make them the way they used to."

Inside, the bakery was stacked to the ceiling with mooncake boxes, barely leaving room for a few racks of puffy breads and pastries. A curt cashier at the counter told me they didn't have lotus-seed-paste mooncakes. I settled for a box of red-bean-paste cakes and sold the other voucher to a ruddy-cheeked young woman who told me she had come to the city with hopes of helping her poor rural family. She paid me 45 *yuan*, but I figured she could probably flip it for an extra 10. At that rate, it sure beat backbreaking farm labor.

"Mooncakes taste horrible, don't they?" the woman said as she counted out the money for me.

"Is that what you're going to tell the person who buys the vouchers from you?" I asked.

She giggled. "They know. Mooncakes are just gifts."

As I walked away, I wondered why I hadn't simply given her the voucher. She needed the money; I didn't. Had living in a country where the official motto of "Serve the people" had been replaced with the unofficial sentiment "Every man for himself" driven me to squeeze every last *yuan* out of a peasant? But with all the bidding and the frenetic pace of Shanghai, the most mercenary of Chinese cities, selling the voucher seemed the most natural thing to do.

After everything I'd heard about the decline in mooncake quality, I no longer wanted to risk spoiling my childhood memories. Later, I met up with a friend who was fretting about not having time to buy a gift for the relatives she was going to see the next morning. I gave her the box of mooncakes. Somebody, I hoped, would eventually eat them.

13

LIVING IN SHANGHAI WAS different this time around. Working three days a week at Whampoa gave me plenty of time to explore. I didn't have to dash around the city, madly trying to interview assorted sources on a myriad of ridiculously different topics so I could meet my deadlines. My social life was quieter too. Many of the people I had known when I lived there previously had moved on, to London, Seoul, New York. I no longer stayed out in bars and nightclubs until the early morning hours. Indeed, without a busy professional and social calendar, I found that Shanghai could be a very lonely city.

One night, I decided to eat at an old haunt, a cheap Cantonese restaurant. I had come for the rice porridge and the roast goose, one of my favorite Cantonese dishes, and also because a casual spot like this was an easier place to eat alone. In China, a restaurant wasn't just a place to eat—it could also be a bar, karaoke club, conference center, or even a wedding hall. If the meal was the occasion for an important business meeting, everyone would stand up and take turns giving toasts until someone at the table fell down. Often a group of friends going out for the evening requested a private room, a common feature of restaurants, which might come equipped with a bathroom, a television, a karaoke machine, and staff to respond to their every demand. Restaurants in China didn't have counters and barstools where you could retreat to eat and read a book on your own. A table for four was usually the smallest available dining space. If I happened to dine alone, I'd feel like a sideshow freak, as waitresses and other diners gave me funny looks and giggled at the pathetic woman with no friends and no family.

That night, I snagged one of the last free tables of no more than a dozen crammed into the fluorescent-lit room with plain white walls. A television hung from one corner of the ceiling, giving the place the feel of a drab hospital cafeteria. A window separated the dining room from the kitchen, where chefs hacked barbecued ducks and geese into bite-sized pieces. No sooner had I sat down than the waiter came by to ask if I would share the space with another solo diner, a woman with a garish, red-tinted bob who hovered before my table.

Sliding into the seat across from me, she ordered noodles and spicy poached chicken. I ordered the goose, my mouth already watering as I thought of the crispy, caramelized skin, the fatty dark meat, and the tangy citrus dipping sauce. When our dishes arrived, we glanced at each other awkwardly and plunged in. A few minutes later, to my surprise, the stranger offered me some of her chicken, and I responded by insisting she try the goose. After a few more minutes, I decided her hairstyle wasn't so bad. It was, after all, the same dye job as my mother's, which made her seem familiar. Halfway into the meal, we were talking as if we were old friends.

She told me she had stopped in for a quick bite before heading home to the suburbs, after a day of shopping in the city. She worked for an American chemical company as an accountant.

"Maybe you've heard of our parent company, Mo Dun?" she asked.

It didn't sound familiar, I told her.

"They make salt."

I thought for a moment. "Oh, you mean Morton! Yes, they are very famous."

She looked at me askance. "That's what they keep telling us at the office. They claim that they control ninety-five percent of the salt market in the United States. But I don't believe that."

I told her that her bosses were probably not exaggerating much. She mentioned that Morton was also making inroads in China. The problem was that Morton salt was too expensive—twice the price of the generic stuff sold in Chinese supermarkets.

"But maybe it's better quality," I said.

"Chinese just care about price. It is true that it's better quality. Morton uses less iodine. But you have to use more of it. I've tried it myself, and it's not as salty as our salt."

I mentioned that in the United States, Morton was the most basic brand of salt. I thought about bringing up kosher salt as an example of the range of salts available there, but I didn't know the Chinese word for kosher. And I wasn't sure why it was called kosher salt anyway. Could I call it "Jewish" salt? I abandoned the idea of explaining any of this, however, when I noticed that my dining partner seemed in shock: "There's salt more expensive than Morton?"

The conversation never strayed far from food, and before long I had enlisted her in my ongoing search for the city's best *xiao long bao.*

"Actually," she said, "I don't think any of the *xiao long bao* in Shanghai are very good. They are better in my hometown,

Wuxi." Wuxi, an hour away by train, had been swallowed up by Shanghai's development and was now considered a suburb. "That's where *xiao long bao* originated."

"Are you sure?" I asked. Everyone I met had a pet theory on this subject.

"Yes, we have rules about making *xiao long bao*. Each dumpling must have sixteen pleats. The Shanghai dumplings are also too salty. Ours are sweeter," she said, as if that settled it. In fact, all the food was sweeter in Wuxi, she said. The food in Shanghai, where locals dumped rice-bowlfuls of sugar into their red-braised pork, wasn't sweet enough for her.

We swapped restaurant tips and debated the merits and flaws of the eateries we both knew. At the end of the meal, she gave me the business card of her favorite restaurant, a place I hadn't heard of. Then, despite my objections, she took out her wallet and insisted on paying the check.

If my social life in Shanghai was a little lonely, I also found it difficult to make friends at Whampoa. The kitchen was divided by language. The overseas Chinese chefs like Jereme and Brother Yao spoke a singsong Cantonese to each other that was as expressive and unique as Italian. The local chefs communicated among themselves in Shanghainese, a language of hard consonants and sharp staccatos that sounded like Japanese spoken by someone on crack. Jereme had also hired a number of deaf chefs through a community service program, and they signed to each other using indecipherable hand motions. The common language was Mandarin, but it was used only to speak with staff members beyond one's own circle. Of all the languages spoken, I could speak just Mandarin, so I was always on the outside.

Anyway, the kitchen buzzed with so much activity that it was hard to get anyone's attention for more than thirty seconds. The only place where people really talked was across the street in Whampoa's cafeteria, where everyone ate before the lunch shift began. The plastic rectangular tables were typical of a Chinese cafeteria, but it had the same breathtaking view Whampoa diners paid top dollar for, of boats and water and the skyscrapers beyond. But even here I hit a wall. When I sat down with a group of waitresses one day, I felt like the new kid in school, all my ice-breaking attempts thwarted. It finally dawned on them that I was the intern. The leader of the pack said, "Oh, right, we got the memo about you." A few minutes later, she picked up her tray and the rest of the waitresses followed, leaving me wondering just what that memo had said.

The cafeteria had its own chefs, who made what I thought was surprisingly decent Shanghainese cuisine. The self-service counter offered a revolving assortment of standards: red-braised tofu, deep-fried perch, roasted pumpkin. Rice, soup, and fresh fruit came with every meal. It wasn't luxurious, but it was far better than what other kitchen workers ate. But the Whampoa chefs had a different perspective on the food, as I learned when I had lunch with two chefs one day.

One of them, a skinny young man who worked in the cutter department, said that as a child he had accompanied his mother to her job at a shipyard, where the food was "much better than this."

The stocky wok chef sitting across from him agreed. "Everything tastes like animal fodder these days. Pigs don't taste the way they used to."

"Pigs used to be individually raised over several years," said the cutter chef. "Now it's done in factories."

"Chickens used to be free-range," the wok chef lamented. "They could wander around and eat whatever they wanted."

"Now they're penned up and force-fed. Meat isn't as good as it used to be." The cutter chef slurped his soup. "But there might be another reason why we think the food used to taste better. We didn't have as many choices back then."

The chefs didn't have a chance to linger over the thought. Abruptly, they stood up, dropped their trays in the dishwashing area, and dashed back to Whampoa. I had eaten only half of my lunch, but I got up as well and hurried after them.

At last a chef befriended me. Her name was Little Han. Just nineteen years old, she handled her job in a male-dominated kitchen with ease. She had been with the restaurant since it opened, having come there right after graduating from a top cooking school.

Little Han was a food caller. Her station was the part of the kitchen that most intrigued me, the culinary equivalent of an airport runway, the counter where the dishes were given a final check before the waiters whisked the plates from the counter into the air above their shoulders and landed them smoothly on the dining tables. Despite their title, food callers didn't do much calling. They were the ones who were called *on*, the "go-betweens for all the sections," Little Han told me. In other Chinese kitchens, she said, the food caller position wasn't that important. But at Whampoa, Little Han played a critical role in giving the restaurant's dishes their theatrical finish, which helped justify the restaurant's high prices.

She had the face of a Japanese anime heroine, broad but with delicate features—small nose, expressive eyes, and narrow lips. She disdained makeup and kept her hair pulled back in a long ponytail, which made her seem even younger than she was. She spoke with her fellow food callers in a low, cracking voice that sounded like a boy going through puberty. And even in her more relaxed moments, she carried herself with a slight masculine shrug. On one of her days off, she stopped by the restaurant in civilian clothes, her hair flowing loosely down her back, and I was surprised that she looked feminine. The professional kitchen made everyone into a man. When I changed into my boxy chef's smock every morning, I noticed the transformation in myself: any hint of my breasts disappeared.

Little Han carried a pen and a knife in one shirtsleeve pocket. In her pants pocket she kept a spoon and a pair of chopsticks. She positioned herself in front of the wok stations. As the wok chefs pushed steaming dishes in tin bowls across the central stainless steel counter, she spiffed up the presentation of the food, transferring it from the dull containers to shiny white plates, martini glasses, and wide geometric bowls. She arranged the ingredients as if prepping them for a photo shoot, applying lines of sauces from squeeze bottles and wiping away errant smudges. She used her chopsticks like pincers, applying sliver-thin garnishes of red pepper or ginger atop the carefully stacked arrangements. She ping-ponged between the various sections, making sure each chef was delivering the necessary items in a timely fashion to maintain the pacing of each meal.

Little Han and I had shared a table in the cafeteria early on in my internship. She had spoken in clipped sentences but was friendly, and at the end of the meal we had entered each other's numbers in our cell phones. She had mentioned her love of *xiao long bao,* for her a close second to the Portuguese egg tart, which despite the name was a Chinese dish—an open-faced tart with a flaky pastry shell and an egg custard filling, torched to give it a slightly burnt finish. She swore by the ones that came from a stand on Huaihai Road. "I didn't know the meaning of happiness until I tried one," she said with a sigh. Then she'd inhaled her food and sprinted back to the kitchen. Since then, we hadn't chatted much, given the demands of her job.

But over time, I got to know the details of her story. She told me she had never thought as a child that she'd end up a chef. "No one in my family is a chef," she said. "I wanted to be a doctor."

She had become interested in cooking after she had developed an intense friendship with another girl. Her friend told Little Han that she liked seafood and street snacks from Beijing, where she had grown up. One snack in particular that she liked was "donkeys rolling in dirt," a glutinous rice cake dusted with soybean powder the color of sand. Impulsively, Little Han told her friend, "I can learn how to make it for you."

"So that's why I became a chef," she told me. "So I could cook for her."

That was at the end of junior high school. Upon graduating, Little Han could elect to go to a traditional high school or a vocational school. In China, traditional high schools were highly competitive and admission was based on the results of a compulsory test. Rather than enter the rat race, Little Han decided to take the vocational school route—it was less prestigious, but it could prove to be more lucrative. It so happened that her grandfather worked for a hotel group that owned the city's best culinary school. At the urging of her friend, Little Han decided to give cooking a shot.

The relationship with her friend didn't last, but her interest in cooking grew. Little Han became an adept chef, and after completing the two-year course, she and eleven of her classmates won prestigious one-year internships at the Whampoa Club. Only a few classmates stayed on after the internship concluded, and two and a half years later, Little Han was the only student who remained. "Some have moved on to other restaurants, but many have left cooking altogether. Everyone has to find an environment they thrive in. The busier I am, the better I do my job," she said.

Whampoa was also the only place she had known—it was her safety zone, just like her parents' home, where she still lived. She had begun to think about leaving the restaurant for better opportunities elsewhere. But tears welled up in her eyes when she thought about moving on. "I know I'll cry the day I leave," she said.

One day, I helped Little Han separate coriander leaves from their stems in the cutter department. It was around ten-thirty in the morning, and there were just a few other chefs scurrying around. In our chef's outfits and paper hats we were practically indistinguishable from the men, but for our ponytails. She said she had been planning to visit a temple, because one of her coworkers in the steamer section was feeling despondent. "He is getting yelled at a lot for doing things wrong. So he wants to go burn incense and pray."

"Does burning incense help?" I asked.

"Well, I don't know," she said. "But at least it doesn't hurt."

As we sorted through the leaves, she asked if I liked to sing. After work, in the late evenings, she and the chefs sometimes went to a karaoke parlor. I said I did, though the truth was, I didn't really care for the way the Chinese did it: they always sang sappy, serious love songs that I couldn't follow because the Chinese words moved too fast at the bottom of the screen.

"I can go, so long as I get home before two in the morning," she said. Her parents worried when she stayed out late. "I have an uncle who is a police officer, and he has lots of stories because he works in homicide."

"Are there a lot of murders in Shanghai?" I asked.

"Downtown is pretty safe. It's the suburbs that are dangerous," she said. I noted that it was the opposite in the States—the suburbs were generally safer than downtown areas.

"Really? How strange," she said. "But even downtown isn't that safe anymore." The previous year, while she was sleeping in her bedroom, on the second floor of her parents' row house, someone climbed in her window, rifled through her belongings, and stole her wallet. By the time she woke up, the thief was gone.

We finished sorting the coriander and headed to the cafeteria, which was serving stir-fried cabbage, red-braised turnips, and a delicious red-braised pork belly steamed with preserved olive leaves. Little Han took care to eat everything on her plate; she'd need the stamina for the day ahead. As with most of the staff, all the trotting around in the kitchen had kept her fit. She had started the job weighing 120 pounds, lost 8 pounds in her first week of work, and after two years on the job was holding steady at 108. "Anyone who wants to lose weight should come work at Whampoa," she joked as she cleared her plate.

My own waistline had been steadily expanding over my weeks in the kitchen. Brother Yao constantly slipped me samples of everything that came off the woks, and I had no willpower to resist. I complained that I was getting fat from all the food.

Little Han laughed and brushed off my comment. And then

I felt silly. There were worse things to complain about, like having to stand on your feet all day and make sure that every plate that left the kitchen was absolutely perfect.

"You have to surround yourself with good people," Jereme told me. He switched to Mandarin to recite a Chinese adage: "Being near red turns you vermilion; being near ink turns you black."

Jereme was talking about the qualities a chef needed to be successful. Just as his cultured army barrack mates had given him a shiny veneer like vermilion, so, he hoped, would the presence of Jean-Georges Vongerichten, or at least his restaurant on the floor below. Now a celebrity, with seventeen restaurants worldwide, including his namesake venture in New York's Trump International Hotel, the French chef jetted around the globe keeping tabs on his various eateries. Jereme mentioned his name in our conversations all the time. "There's a learning process going from cook to entrepreneur. Jean-Georges tells me this," he said one morning in his office. Another time, as if confessing a deep secret, "The more I look at it, I'm not suitable to be a Jean-Georges."

One October day, the French chef was passing through town and decided to have lunch at the Whampoa Club. "Jean-Georges—VIP" was the scrawled heading on the memo a waitress distributed around the kitchen at noon, shouting like a town crier, "Eight VIPs at two-thirty!" The kitchen was already running on overdrive to deal with a luncheon for thirty. Tommy, a senior waiter whose narrow eyes reminded me of a jackal's, exploded into rapid-fire Shanghainese, impatient for a set of cold dishes to materialize. The cold-dish chef hunched over a set of plates and worked at his usual measured speed, unfazed by Tommy's outburst.

The kitchen was further handicapped by the absence of Lim, the new sous-chef, who had taken the day off to pose for wedding photos with his new wife — in China, a full-day affair. So rather than pacing around the kitchen, Brother Yao was anchored at a wok station. It was one of the few times I had seen him at the wok, and I was struck by the economy of his movements. He flipped a shallow tin bowl of ingredients upside down, and they went tumbling into the wok. He snapped his spatula against the bowl to release anything that stuck to the sides and sent it spinning like a Frisbee across the counter until it fell into a crate of dirty tins. He slammed the spatula against the wok and whisked the ingredients vigorously over a large fire. He picked up the wok, gave it a few shakes, and turned his body 180 degrees to face the counter behind him. Bending his knees slightly to bring his waist level with the counter, he scooped the contents of the wok into a clean tin bowl and yelled for a food caller. The process took no more than a minute.

In the dim sum room, the chefs glanced at the memo that had been dropped on the counter. It listed the dishes Jereme had planned for his distinguished guest, a menu that highlighted Jereme's pan-Chinese influences:

> *Stir-fried hairy-crab meat and ginger—egg white custard*
> *Crispy egg noodles with hairy-crab and pork dumplings*
>   *infused with aged vinegar*
> *Steamed crystal dumplings with sweet shrimp*
> *Shrimp and pork shaomai dumplings with air-dried*
>   *duck liver*
> *Long seafood spring rolls*
> *Black sesame and green bean jelly, flavored with coconut*

Spicy rabbit-shaped buns with five-spice pork (savory)
Cute silkworm-shaped dumplings with shrimp and
    almonds (savory)
Mini savory pancakes, Cantonese style

The trio of dim sum chefs usually worked at a leisurely pace, making lewd jokes as they went. ("I haven't done it in weeks," I heard one chef say to another, and even with my feeble Cantonese, I was pretty sure what "it" was.) But today there was no joking; all four had their heads down as they concentrated on folding pastries into intricate shapes at top speed. The chefs placed the folded pastries helter-skelter in a steamer basket, like a cage full of rabbits, until one chef organized them into a neat, inward-facing circle of well-behaved marshmallow Peeps.

Yap, a portly chef with flaring nostrils who ran the dim sum department, poked his head into the alcove. "How are the *har gaw?*" he asked, referring to the shrimp dumplings.

"Ready," a chef replied.

"Be ready in ten," Yap said, hurrying off.

Outside the dim sum room, chefs swarmed like bees, shoving and bumping each other. Dropping the name of the head chef of French Laundry and Per Se fame, Jereme had once told me that a "Thomas Keller kitchen works like a Swiss watch." In European or American kitchens, he'd explained, it was possible for chefs to work individually, without much communication. This was not so in Chinese kitchens, even the most vaunted ones, with all the various ingredients that had to be prepared and served immediately. At Whampoa, everyone was in everyone else's way. The crashing sounds of spatulas banging on woks intensified like cymbals at a symphony's climax. A battery of tins went spinning into the dirty dish crate. Everyone was shouting.

The chefs and waiters: "Coming, coming, coming!" "Move, move, move!" "Hurry, hurry, hurry!"

"Gently, gently!" Brother Yao growled at a chef carelessly tossing expensive plates onto the counter.

The steamer department head to a deaf chef doing his best to lip-read: "Parsley, parsley!"

Tommy, the jackal-like waiter: "*Nie, nie, nie!*" *Nie* didn't mean anything at all, but was used to get someone's attention.

Waiters, carrying trays covered with clear domes that kept the contents warm, propelled themselves through the swinging doors and into the dining room with their backs. Waitresses barreled through from the opposite direction, carrying empty trays ready to be reloaded. A stack of clean white plates went one way, a bowl of chopped scallions the other. A clutch of hairy crabs shot by at dizzying speed, a bulbous mass of ginger, a brace of roast ducks.

From the dessert room, the smell of puréed banana drifted into the hall. The dessert chefs arrived after everyone else but had to stay late into the evening, since their dishes came at the end of the meal. As the rest of the kitchen was engaged in the lunch-hour commotion, they worked in relative peace, in a haven of freshly grated ginger and vanilla and other blissful scents, which wafted through the kitchen until the ventilation system sucked them out, along with the heat and the grease.

Although Whampoa's was the cleanest kitchen I had ever seen in China, it often failed to meet the standards of Dr. Chen, a health inspector who spent her days tramping through every restaurant kitchen in the building. That afternoon, she marched through our kitchen, reading glasses in one hand, a plastic file of checklists in the other, a weary expression on her face. She looked like a Chinese version of Nancy Reagan, her white blouse tailored, her skirt long and dark, her hair short and wavy. Her ongoing campaign to reform the kitchen's sloppy habits was often met with the same derision that had greeted the former First Lady's "Just say no" campaign.

"The fans aren't on high! They should be on high!" Dr. Chen wailed. She rattled off her complaints: the chefs didn't wash the eggs before they cracked them into dessert batters. The rack of plated desserts was not supposed to be sitting next to the sink. The chefs went to the bathroom and didn't wash their hands! Why was there a pile of raw ducks near the appetizers? Some of the chefs were wearing rings!

"I've told them already not to wear rings!" she exclaimed. The chefs gave her bemused looks and went about their business. She turned to me, gasping: "Help me out here! Tell them for me, will you?"

"I'm busy. I can't listen to her all the time," said a dessert chef, as if he and Dr. Chen were in a family counseling session and I was the psychologist. The chefs referred to Dr. Chen as "Auntie." He knew that it was Auntie's job, "but if we listened to her all the time, we wouldn't be in business anymore."

Dr. Chen left with a huff, off to another kitchen on the floor above or below, where she would likely receive the same treatment. The final trays of desserts went out to the lunch event, and half the chefs departed. The other half remained in the kitchen, waiting for the VIP party of eight, pacing and fussing like expectant fathers. Dumpling pleats were scrutinized, woks were brushed, ingredients were queued.

At two-forty, a waitress popped her head into the kitchen: Jean-Georges had arrived. Everyone sprang into action.

A chef dribbled sauce made from hairy-crab meat onto egg white custards that had been steamed in hollowed-out eggshells. A sprig of scallion was inserted into each shell.

Brother Yao's cell phone rang. Still hunched over an egg, he answered, yelled "Okay, okay!" and hung up. It was Jereme calling from the dining room, demanding to know where the food was. It surprised me that he wasn't in the kitchen that day, but I guessed he had calculated that it was more important for him to be with his venerated guest.

In the kitchen, the chefs tossed silkworm-shaped dumplings with paper-thin pleats of dough into the wok for a scalding oil bath, then arranged the crisped dumplings in a dim sum basket. Using a toothpick, a chef attempted to apply individual sesame seeds on each side of the fried dumplings, to give them "eyes," but the seeds were being temperamental. A few other chefs joined in to help, and soon, four men stood with toothpicks in their hands, trying to stick the little black seeds onto the cute, pudgy dumplings.

Egg rolls shaped like batons and coated at each end with sesame seeds emerged from the deep fryer. Someone arranged them in cocktail glasses as if they were flowers, stuffed lettuce leaves into the glass like tissue paper in a gift bag, and trickled extra-thick soy sauce atop the leaves.

A chef injected aged vinegar from a syringe into each of the hairy-crab and pork dumplings as they grilled in a skillet. The crispy pancakes were similarly pan-fried, spread with a mixture of chopped Chinese leeks and pork, folded, and cut into four sections that were then arranged on a plate like a pyramid and sprinkled with strips of nori. Hot off the steamer, the bamboo baskets were opened and the buns and dumplings inside were brushed with sesame oil. As each dish was finished, a perspiring chef whisked it off to the VIP table.

Several chefs ushered the final dish into the dining room just before three o'clock. The dozen remaining chefs unclenched and pounced on the leftovers.

After two years in the Singaporean army, Jereme returned to cooking. He bounced around the Cantonese kitchens of various four-star hotels before breaking into a five-star resort, the Mandarin Oriental in Surabaya, Indonesia's second most populous city. The resort put him in charge of Sarkies, the property's Chinese restaurant.

A Chinese kitchen at a five-star hotel generally needed three key chefs—a dim sum chef, a wok chef, and a general operations chef, Jereme said. To save money, the hotel hired him to do all three jobs. The owner of the property was a businessman who also ran a frozen-food business. The posting—in addition to doing a job meant for three—required Jereme to work with the frozen-food company. When he wasn't cooking in the kitchen, he sourced ingredients and worked with food chemists to devise dim sum dishes that would freeze well.

Somehow, on top of all his duties, Jereme managed to make time to experiment with food. He added goose liver to his *xiao long bao,* and fish roe to his pork *shaomai.* He took a sweet Chinese dish of *guilin gao*—a gelatinous herbal jelly—and deep-fried it. "Usually, if you tried to deep-fry it, the gelatin would make it melt," he said. "So I replaced the gelatin with flour and agar-agar," a seaweed-based ingredient that helps liquids set.

Sarkies, under Jereme's leadership, became known as one of the best restaurants in Surabaya. With a 180-seat dining room,

"I was doing six hundred covers at lunch," Jereme said. People waited for up to two and a half hours to eat. "You were really a king in those days," he said, referring to himself in the second person. "I mean, I was really" — he scratched the tip of his nose with his finger — "in those days."

In the years before the Asian financial crisis of 1997, the luxury market in Asia boomed, and the Mandarin Oriental sent Jereme to oversee the launch of its Chinese restaurants that opened with each new property the hotel chain built. Jereme jetted from Jakarta to Hong Kong to Manila to Kuala Lumpur, a man in his twenties bossing around chefs in their sixties. In Asian cultures, where age is associated with authority, the arrangement sometimes caused tension. "My attitude was not 'you are wrong' but 'this is wrong.' In any operation, if you allow people to do what they want, it doesn't work," he said.

At the Mandarin Oriental, Jereme also came in contact with a number of visiting European chefs, who helped him cultivate a taste for Western delicacies like the foie gras he had so disliked. The various hotels organized round-table luncheons featuring foreign cuisines made by visiting chefs. "One day it was French, then Indian, then Chinese," Jereme said. "The French chef would give us really stinky cheese. And I would think, yesterday it was cheese, today I'm going to make him eat durian pancakes!" (A popular fruit in Southeast Asia, durian has a reputation for being the smelliest fruit in the world. It tastes and has a texture like a cross between an onion and a pineapple but reeks of dirty socks from yards away.)

By 2000, when he was just twenty-nine, Jereme had made a name for himself in the Asian hotel trade. His durian pancakes had been a hit among Malaysian diners at the Mandarin Oriental in Kuala Lumpur. He had received prestigious awards, but he felt as if he were stagnating. The Mandarin Oriental offered to transfer him to the corporate side and train him in "F and B"—the food and beverage department. "They wanted to send me to Lausanne and come back and do the same shit, except I would be a manager," he said, referring to the Swiss town's well-known school of hotel management.

So when the Four Seasons Hotel in Singapore approached him, he decided to return to his adopted homeland. "I was back in the pool, in a more developed place. That offered another level of challenge." The challenge wasn't in the kitchen —he felt he had already refined his concept of food by that point—but in learning how to deal with customers. At the Mandarin Oriental, "I didn't care about meeting guests. The Four Seasons operated differently. If things were wrong, you get to tell me that my latte is shit. You would hear it firsthand."

One day, three diners approached Jereme after sampling his food. They represented a group of Asian investors who were renovating a historic building on Shanghai's riverfront into a retail and dining complex that would be called Three on the Bund. The investors were looking for a chef who could do classic Shanghainese cuisine in a fine dining environment.

One of the three representatives was Handel Lee, the cochairman of Three on the Bund and a Chinese-American lawyer. I spoke with Lee about what he thought Jereme offered. The group couldn't find a Shanghainese chef who could execute what they wanted, he said. "The chef would need foreign experience from a five-star hotel. The chef needed to meet a standard of cleanliness, presentation." After considering more than a dozen chefs from around Asia, the Three on the Bund executives selected Jereme. Among the factors that earned him the job, Lee said, was the fact that Jereme was "so *jiepi*"—essentially, the Chinese way of saying "anal-retentive." "All his dishes were stunning, plated beautifully. We saw his kitchen. It was so clean and orderly. He was a drill sergeant."

Jereme was inclined to accept the job after he heard that "Jean-Georges would be downstairs." He also liked the idea that he would have the freedom to create dishes of his choosing and wouldn't have to jump through bureaucratic hoops to get changes approved, as at the hotels where he had worked. He liked being able to fire whomever he chose, whenever he chose.

It wasn't a handicap that Jereme "knew nothing about Shanghainese food," Lee added. The restaurant scene in Shanghai was so crowded with places trying to do "nostalgic Shanghainese" or a mishmash of cuisines that Jereme's lack of familiarity with the city was considered an asset, Lee said. Three on the Bund hired him to move to Shanghai a year ahead of the restaurant's opening. He researched the local cuisine and took cooking lessons from a number of traditional chefs, octogenarians who taught at a respected cooking school in Shanghai.

Before Jereme came to mainland China, he had turned up his nose at its cuisine, an attitude that was shared by many Hong Kong and overseas Chinese. He thought that "Chinese food was Hong Kong food. Other cuisines were just Chinese food on the side." Moving to Shanghai "really opened my eyes to the history and culture. I realized how much I didn't know about Chinese food. Before coming, I wouldn't have touched pickled vegetables. I would think of the dirty container they came in. But then you realized when you see it carefully that it's sealed up and the fermentation makes it clean."

Jereme expounded a little on how he had revitalized the

Shanghai smoked tea egg, a dying classic, by adding caviar to it. "We made it a big hoo-ha. Because of a twist, you've brought it back from the dead." The dish was now copied everywhere in Shanghai, he noted. "It makes me feel good. You laugh and say, Copycat!" He giggled uproariously. But he also didn't hide his impatience with the research period that preceded the restaurant's opening, which was extended because of construction delays and the SARS epidemic of 2003. "The restaurant could have opened earlier." And by the shake of his head, I gathered he meant *a lot* earlier.

The evening after Jean-Georges came to lunch, I returned to Whampoa to see if I could pull off an entire day in the kitchen without collapsing. I spotted Jereme in his chef's whites in the steamer section. In recent months, seeing the founding chef in uniform had become as rare as spotting a panda in the wild. But there he was, his thumb grazing his chin, his eyes set somewhere in the distance. The sleeves of his jacket were rolled up around his elbows. The white smock appeared to be a size too big for him, as though he didn't quite fit into his environs.

Now that things at the Whampoa Club had run smoothly for two years, Jereme was setting his sights on new projects. He had been flying back and forth to Beijing, where he planned to open a second Whampoa Club in the next few months. He had also been zipping around Asia. Despite having a full-time job at Whampoa, he was taking assignments from his former employer, making trips to Kuala Lumpur and Bangkok to oversee events and redesign menus for the Mandarin Oriental. This sort of arrangement — businessmen juggling projects for multiple employers with no perceived conflict of interest — was typical in Asia.

When Jereme was in Shanghai, he spent much of his time in his office: on the phone, answering and writing e-mail, holding meetings with the kitchen staff. Once in a while, he strode through the kitchen, and the chefs—particularly the more junior ones—flinched as he approached, as if they had suddenly contracted a nervous twitch. Jereme rarely yelled or threw temper tantrums. When he scolded, it was in an even-pitched and controlled manner—but it was relentless. Little Han recounted how a staffer assigned the task of asking Jereme what he wanted for lunch had come to Little Han in a panic, begging her to do the job for him. Fine, she said, wondering what the big deal was. But as she approached the office door, something she saw through the window made her cower.

"I was so scared," she said. "I ran away and tried to get someone else to do it."

"Why?" I asked.

"Jereme was in there scolding one of us," she replied. "I couldn't see Jereme's face, but I could see the chef's. His face had turned completely white."

Approaching him now, I felt as frightened as his staff. Who was I, a lowly intern, to approach him in his kitchen?

I feebly asked how the lunch with Jean-Georges had gone.

Jereme's gaze was focused elsewhere, and I felt as small as a fly. Finally he responded. "That guy is opening far too many restaurants." I didn't have a chance to decide if the remark was tinged with envy or admiration before he hurried off to the other end of the kitchen.

14

ONE EVENING, I STOPPED in at a restaurant named Yin that I had frequented when I had lived in Shanghai. Located in a grassy compound that contained some of the city's most impressive historic buildings, the restaurant felt like a simple, elegant home. A small foyer led to a large dining room with wood floors, contemporary paintings, and antique Chinese furnishings. The food was similarly understated—small servings of modern Shanghainese and Sichuanese dishes cooked by the same chef, night after night. Yin was part of the movement to reinvent Shanghainese food, if on a humbler scale than the Whampoa Club.

Chef Dan served his chicken and fish boned. He didn't use MSG and instead flavored his dishes with a chicken and pork broth. He used oil sparingly. His style was unpretentious: he used few exotic ingredients, toned down his spicy *douban* chili sauce with sugar for spice-adverse taste buds, and smoothed out his Sichuan-style *dan dan* noodles with peanut butter—concessions that would have been considered repugnant to a traditional Chinese chef.

A travel magazine editor I knew once noted, "There is a difference between best restaurants and favorite restaurants." To my mind, this seemed to be the difference between the Whampoa Club and Yin. Whampoa was striving to be the pinnacle of Chinese cuisine, a place where diners went for a special experience; many considered it the best restaurant in Shanghai. Yin was a neighborhood restaurant full of regulars who considered it their favorite.

As always, Yin's owner, a Japanese man named Takashi Miyanaka, stood in the foyer greeting his guests. Exceptionally polite and soft-spoken, Takashi tilted his head as I stepped inside and removed my autumn coat, his wispy locks grazing his shoulder, a surprised smile on his face. I had gotten to know Takashi and Chef Dan when I lived in Shanghai, but it had been more than two years since I had last visited.

As I sat down to eat, Chef Dan emerged from the kitchen to greet me, wearing a black mandarin-collar shirt, jeans, and tortoiseshell glasses. He had a wrinkly forehead, a bony frame, and a complexion the color of a long-brewed oolong tea. He smiled and nodded, just as timidly as Takashi. His self-effacing demeanor reminded me of one of those middle-aged Shanghainese men I saw in the parks who took their caged birds for walks. But when I told him that I was back in Shanghai to learn to cook, his face lit up, and he reverted to his lively kitchen personality. "Once you have the basics down, you can apply them to anything!" he said. "For example, when I was working in a restaurant in Guizhou" — in southwestern China — "the boss came in with a live civet cat."

"Aren't civets banned now?" I asked. The raccoon-like animal had been linked to the SARS epidemic.

"This was long before. Anyway, I had never eaten civet cat. I'd never prepared dog or cat, either. So what did I do? Well, first I shaved it. It was a ferocious animal." He scowled, trying to convey the feral nature of the beast. "Then I threw it in a vat of boiling water." He mimed pitching it across a roomful of diners. "Then I skinned it. And then I braised it. Soy sauce and sugar. It tasted great." The lesson: "You shouldn't say you can't do something just because you haven't seen it done before." He invited me to visit him in the kitchen anytime to work on my skills.

From time to time I took Chef Dan up on his offer. The kitchen—a simple room with two wok stations, a large counter, and two sinks—was a pleasant counterpoint to Whampoa's intensity and obsessive attention to detail. One afternoon I found Dan at the wok. He glanced in my direction and hollered above the whir of the fans, "Why don't you make the boss's lunch?" He called for a waitress, who rushed over with an apron and a paper chef's hat. Takashi had requested his usual: home-style tofu with rice.

"Go easy on the spice," Dan said. "He doesn't like it very spicy."

Once, Jereme had let me try out Whampoa's wok stations, with the supervision of one of his senior chefs. But under no circumstances, he told the chef, was I to cook anything that would be served to the public.

I assumed that Takashi would forgive Dan if the dish was a disaster, but cooking for the boss made me nervous. Dan didn't seem to notice. He instructed me to coat the tofu slices with flour before deep-frying them. When they were crisp and browned, I drained them. Into a clean wok went oil, scallions, and ginger. Then I shook a dollop of broadbean paste and a dab of chili sauce into the wok. "Add the tofu!" Dan commanded over the roar of the fan. The tofu sizzled in the wok. I poured a shot of rice wine over the mixture, allowed it to bubble and simmer for a minute or two, then added soy sauce and reduced it some more.

Once I began cooking, my instincts kicked in. I finished the dish confidently and transferred it to a glazed porcelain plate, which the waitress carried away. Watching the dish go from cutting board to wok to table, I felt the same sense of accomplishment I'd had when I successfully grated a bowl of noodles in Chef Zhang's stall. I had made food, and somebody other than an effusive friend or family member was going to eat and judge it.

On my way out, I ran into Takashi. "How'd you like the tofu?" I asked him.

It took the handsome owner a moment to respond. "You— you made it?"

HOME-STYLE TOFU (*JIACHANG DOUFU*)

1	12-ounce package firm tofu
½	cup all-purpose flour
¼	cup broadbean paste
¼	cup Chinese chili sauce
1	teaspoon sugar
1	cup vegetable oil
1	tablespoon chopped scallion
2	thin slices ginger
1	tablespoon Shaoxing rice wine
2	teaspoons soy sauce

Cut the tofu block in half lengthwise, then crosswise into ¼-inch-thick slices. Pour the flour into a small bowl. Dredge each piece of tofu in the flour and arrange on a plate in a single layer. In another small bowl, mix together the broadbean paste, chili sauce, and sugar. Set the mixture next to the stove.

Place ¾ cup of the oil in a wok over high heat. When the oil is hot, add the tofu slices one by one, arranging them so their undersides are covered with oil. (You may have to fry them in two or three batches.) When one side is lightly browned, turn the pieces over and brown the other side. Remove the tofu from the wok and put the pieces on a clean plate.

Pour off the used oil and pour the remaining ¼ cup of oil into the wok and place it over high heat for 1 minute. Add the scallion and ginger and stir for 1 minute. Add the broad-bean paste mixture and stir for another minute. Add the tofu, rice wine, and soy sauce. Reduce the heat to medium and simmer for 3 to 4 minutes. Remove from the heat and serve immediately.

◎

Tap-tap tap-tap went Dan's ladle as it danced in the wok. He grasped the wok with a wet cloth and gave it a deft shake, sending bits of pigeon, pine nuts, and diced shiitake mushrooms into the air before landing soundly back in the iron pan. Next to the wok station, his sauces and spices were arrayed in neat containers that looked like paint ready to be applied to a canvas.

Off to the side I could see Dan's sous-chef tossing *shizitou*, balls of deep-fried pork and crab, into a vat of boiling water before stewing them in soy sauce. Across the kitchen, an assistant sliced a beef tongue. Another two assistants stood facing Dan at a prep counter, waiting for each dish to come off the wok. Yin had the intimacy of a tight-knit family operation.

I felt welcome in the kitchen at Yin, and at ease around Dan. I'd felt comfortable enough with him to reveal my Taiwanese roots. After the lunch shift was over, we sometimes rested in the dining room or on the patio and talked about food, and I jotted down a few of Dan's comments:

In China, business people go to restaurants to talk about business. No one actually talks about business in the office.

There's no talent involved in cooking.

I think foreigners are not so interested in eating *real* Chinese food.

One afternoon as we sat in the darkened dining room, Dan talked me through the preparation of red-braised pork.

"Buy two *jin* of pork belly. You need a good amount of meat, at least two *jin*. It's like making rice. If you make too little, it won't come out right. Cut it into cubes. Boil it in water for five minutes." He paused and took a puff of his cigarette. "Drain it and wash it under cold water to get rid of the impurities. Cut up some sections of scallion and ginger and stir-fry them in oil. Add the meat and sauté for a few minutes. Add some rice wine, about three porcelain spoons' worth. Use a big fire!" He paused and took a sip of his coffee before he continued.

Until I met Dan, I had never known a Chinese chef who liked coffee. Chefs always drank tea out of large glass tumblers, which they refilled with hot water from time to time. He'd told me that he'd picked up the habit from his father, a successful architect who had learned to drink coffee during his studies abroad. Coffee wasn't hard to come by in Shanghai. Before the Communists came to power, the city had been home to a British-American joint-venture coffee roaster. When the foreigners left, the government took over and continued producing the beans. Dan's father didn't fare well under the Communists. Being an architect was bad enough, but some of his father's relatives had sided with the Nationalist government. They had fled to Taiwan in 1949 to escape persecution, but Dan's father didn't make it out in time, and "life got very difficult for him," Dan said with a nervous chuckle. Still, the bourgeois coffee habit stuck, and Dan's father passed it down to Dan. When he'd first gotten a job as a chef, at eighteen, he'd spent 40 cents of his monthly salary of $4 on a bag of freshly ground coffee, slowly savoring it until his next paycheck.

The government had placed him at a state-owned restaurant called Meilongzhen. Though the restaurant was one of Shanghai's best, Dan wasn't particularly inclined to be a cook. "I didn't even know how to hold a knife," he said. He started by working as a vegetable washer and slowly worked his way up. He ended up staying at the restaurant for eighteen years, and from what little he said about that time, it sounded dreary. In the 1970s, few people could afford to eat in such an establishment, and delicacies like shark's fin were unavailable. It was only on certain state-sponsored occasions, like when Cultural Revolution stage performers passed through town, that the restaurant came to life.

In the 1980s, however, foreign traffic picked up. And in 1991, when Dan was a senior chef, a German restaurateur came to dine at Meilongzhen. He liked Dan's food and offered him a job at one of his restaurants, in Bonn. Such offers were rare at the time, and with the restaurateur's sponsorship, Dan was able to get a passport and work visa. In Bonn, he cooked by day and slept on a tatami mat in a small, unfurnished apartment at night. His salary at Meilongzhen had topped out at $40 a month; in Germany, he started at $1,200 a month. The restaurant wasn't to his liking, however.

"I was surprised when I got to Germany," said Dan. "I went there and discovered, 'It's like this?'" A long, pregnant pause. In Germany at that time, "the Chinese restaurants had no *xiao long bao*. They didn't know what fish-fragrant sauce was. They didn't know the right ingredients for Yangzhou fried rice. They didn't know the difference between Chinese food and Japanese food." So that was what accounted for Dan's musing about how foreigners weren't interested in eating real Chinese food.

In Bonn, Dan ran into an old chef acquaintance from Shanghai who had also moved to Germany. The friend introduced him to a network of Chinese restaurants and chefs, and Dan moved between various eateries all over Germany. After three years abroad, he decided to go home. "I liked Europe. But the problem was, I couldn't stay in Germany." He paused. "Well, I could have. At the time, Europe was allowing Chinese to stay." Since the Tian'anmen Square massacre in 1989, Chinese could plead for political asylum. But by taking that route, he wouldn't have been able to return to China, at least for the foreseeable future, and he had his mother to think about.

After returning to China, Dan traveled around the country for several months, venturing far into the hinterlands, his German savings giving him the freedom to explore. In the Muslim province of Xinjiang, which borders central Asia, he developed a taste for cumin and tomatoes. He worked briefly in a restaurant serving Shanghainese cuisine in the poor, mountainous province of Guizhou, where he had skinned and red-braised the fierce civet. He went to the southwestern province of Yunnan, where he ran a restaurant owned by a tobacco company. Then he returned to Shanghai for good.

Not long after coming home, he met Takashi Miyanaka, who wanted to open a nontraditional Shanghainese restaurant. It would be the kind of place where you could start with a glass of Spanish wine and end with a cappuccino, and in between enjoy a good meal in a peaceful setting. Dan was just the kind of chef that Takashi was looking for—someone who had experience but wasn't set in his ways.

Takashi let Dan take control of the menu. Dan drew on his Shanghainese childhood, his years of Sichuan cooking at Meilongzhen, and his experiences traveling around China to create dishes like "Mr. Dan's Special Sliced Lamb and Garlic Sautéed with Ten Different Exotic Spices."

I once brought Teacher Jiang, my food critic friend, to eat at Yin. After lunch, he fanned himself and looked around at the elegant surroundings. "You want really good Shanghainese food?" he sniffed. "I'll take you somewhere else." I had met Shanghainese who had a downright hostile attitude toward Yin. They called it "fake" Chinese food. It was too simple, too home style. At the prices Yin was charging—a meal cost between $10 and $25 per person—they expected fish head, at least. Maybe even some shark's fin or abalone.

At Yin, I realized that the idea of food being "authentic" was relative. Here I was in Shanghai, eating Shanghainese food made by a Shanghainese chef, and some people still didn't consider it the real thing. What exactly was "authentic" Shanghainese food? Was it the basic food that the Shanghainese ate before it became a booming port, when it was largely devoid of foreign influence? Or was it *benbang cai,* the bastardized fusion that Shanghai became known for during its colonial era? Or was it Jereme's novel dishes? When people talked of "authentic," they didn't take into account that food was always changing and adapting, a reflection that the place it was from was also changing too.

And I liked that Yin felt a bit more like home, with its cushy interior, its wine selection, and its unfussily presented, simple dishes. I stopped feeling embarrassed about liking Yin as I knew more about "real" Chinese food. Learning how to eat a foreign cuisine was like learning a foreign language. It took years to do it, and even after becoming fluent, it didn't mean that I always preferred the Chinese way of eating or speaking. Sometime during my stay in Shanghai, I realized that my taste buds—just like my personality, my outlook on life, and my political views —had been shaped by my childhood in America. While they had been modified by my years in China, they would remain fundamentally unchanged. I was happy being who I was, whatever that was. Perhaps I was simply an American who was interested in —as gruff Teacher Zhang had put it early on—my "roots."

RED-BRAISED PORK (*HONG SHAO ROU*)

2	pounds pork belly, cut into 1-inch cubes
½	cup vegetable oil
1	tablespoon chopped scallion
4	thin slices ginger
2	tablespoons Shaoxing rice wine
½	cup dark soy sauce
⅓	cup sugar
2½	cups water

Fill a medium stockpot halfway with water and bring to a boil. Add the pork belly cubes and cook for 5 minutes. With a spoon, skim off any froth. Drain the pork and rinse under cold water.

Place the oil in a wok over high heat. When it is hot, add the scallion and ginger and stir for 1 minute. Add the pork cubes and stir for 3 to 4 minutes. Add the rice wine, stir for 1 minute, then add the soy sauce and stir for another minute. Add the sugar and reduce the heat to medium, stirring for 1 minute. Add the water, then cover and simmer over low heat for 2 hours. Serve over white rice.

◎

Truth be told, the *xiao long bao* I ate at Whampoa weren't the best I'd ever had. The skins were light and tender, the meat was high quality, and the soup was laboriously made, but there was something lacking, a certain *je ne sais quoi*. Jereme told me that he wasn't trying to beat the locals at their own game: "When I go to Beijing, I'm not going to try to make a better Peking duck. You focus on doing other things differently." The Beijingers had been at it for decades, if not centuries. The same applied to the *xiao long bao*.

So I continued my search for the perfect Shanghai soup dumpling. Teacher Jiang offered to introduce me to the manager of Lü Bo Lang, a traditional Shanghainese restaurant that sold more than ten thousand *xiao long bao* a day. Located in a touristy part of town, the restaurant had originally been an exclusive dining hall for Communist Party members. It had opened to the public in 1979, after the economic reforms began. As one of the state-owned restaurants most favored by the government, Lü Bo Lang often received foreign dignitaries. Teacher Jiang mentioned that Bill Clinton had once visited. "I told the manager that he should have saved the chopsticks Ke-Lin-Dun used. They would have been worth a lot of money!"

The restaurant was a narrow, three-story building with an entrance that resembled a Chinese pavilion. The sloping roofs had pointed corners that tapered and curved upward, Ming Dynasty style, though the building had been built just decades before. It sat next to Shanghai's biggest tourist trap, Yu Yuan, which was a poorly maintained Ming-era garden. But that didn't stop visitors from arriving by the busload. The area surrounding the garden teemed with Chinese and Western tourists, who wandered a narrow labyrinth of walkways lined with shops selling imitation jade and synthetic silk purses.

I was skeptical when I arrived—the closer a restaurant was to a tourist hub and to the government, the worse it usually was. But I suspended judgment because I respected Teacher Jiang's opinion.

At the restaurant, the manager said I could wrap a few dumplings in the pastry room before I tried them, but he would have to get me into the kitchen covertly, after the lunch rush ebbed; the health inspectors happened to be visiting. While we waited, the manager recounted the story of a visit by the exiled Cambodian prince in the early 1970s. The prince had fled to China after being forced out of his country. A friend of Chairman Mao's, he received lavish treatment when he arrived in Shanghai. Government officials notified Lü Bo Lang that the prince was planning to stop by for a meal—but they didn't say when. For three days, the entire area was sealed off as everyone waited for him to arrive. On the third day, after playing a round of golf, the prince finally showed up, sending the chefs into a panic. The Party leaders had commanded the kitchen staff to make a special soup of duck's blood and ovaries of a particular size and shape. The chefs slaughtered two hundred ducks to find the perfect specimens. Lü Bo Lang didn't host VIPs that way anymore, the manager said.

In the dumpling room, a dozen middle-aged workers sitting on barstools hunched over a counter assembling the *xiao long bao*. It looked like the most boring job a cook could have, worse than working at the dumpling factory. The Beijing dumpling restaurant had used a variety of meats and vegetables. But the *xiao long bao* had the same stuffing, very fatty ground pork. I was surprised to find that the making of the dumplings involved no rolling pins—the chefs smushed the wrappers with their palms and quickly smeared the filling into their centers before finishing with a few cursory folds. On a busy day, two workers could wrap 3,600 *xiao long bao*.

After I wrapped a tiny fraction of the day's dumplings, one of the workers brought a basket of them for me to try. They were some of the worst I had ever had—the skins were thick and gooey and the meat tasted like something from a can.

A few days later, I went in search of the birthplace of the *xiao long bao*, an expedition that took me by bus along the elevated highway to Nanxiang, the landscape getting grittier by the minute as the bus drifted away from downtown. For the average Shanghainese, spoiled by having all the conveniences of life within a ten-block radius, going to the suburbs was as dreadful a proposition as a trip to the outer boroughs is for the bona fide Manhattanite. Even Teacher Jiang, the most adventurous food critic I knew, had never been to Nanxiang. When I'd asked him if he wanted to come along, he'd said, "No, not really," and wished me luck.

I wasn't even sure I was going to the right place. Locals had given me different answers when I asked them where the *xiao long bao* originated. The most plausible place, I gathered, was Nanxiang, because there was a historic dumpling shop in Shanghai with the same name, and Teacher Jiang, who had a better grasp of Chinese history than anyone I knew, had been insistent that Nanxiang was indeed the place.

Overloaded trucks covered with tarps rumbled by. Out the window, endless rows of ugly housing units whizzed past. A layer of haze kicked up by speeding cars hovered above the bumpy roads. A man with a dirty face sat near a highway off-ramp holding a sign that read DISSATISFIED.

The bus passed a rubble-strewn lot and an abandoned factory with a smokestack rising behind it. "Nanxiang," said the driver. He might as well have said "the moon." I climbed down to find men lazily lounging on motorbikes, hoping to shuttle strangers for exorbitant fees. The drab avenue was lined with shops selling an assortment of construction supplies. A speaker in front of a shoe store shrilly advertised the store's pleather loafers. There was no dumpling shop in sight.

I hailed the first taxi I saw and asked the driver if I had gotten it right: was this Nanxiang, the home of *xiao long bao*? He drove me to a cluster of dumpling shops a few blocks away. Dozens of shabby restaurants on an alley advertised themselves as the source of the authentic *xiao long bao*. I did a brief count of the number of stalls and contemplated how many dumplings I could stomach. Then I came across a more official-looking restaurant called Gu Yi Yuan, next to a well-kept garden, that looked more promising.

Though it was housed in an old Chinese mansion, the interior of the restaurant felt like a canteen: the drafty, high-ceilinged dining room was decorated with blue ceiling fans, simple tables and benches, and scowling waitresses. The *xiao long bao* at Gu Yi Yuan were as horrendous as the neighborhood. The skin was thick and spongy, and the meat tasted like rubber. After swallowing a few, I decided I was done.

A brief interview with the manager didn't clear up the history of the *xiao long bao*, either. The manager claimed that a steamed-bun maker named Huang Minxiang invented the dumpling in 1871, to give him an edge over his steamed-bun competitors. But he didn't have any historical documents to back up his claim, and he was reticent as I pestered him with questions about long-ago times. He was more interested in the future. A new subway line was being built from downtown

Shanghai that he hoped would improve his business. The State Council had designated the restaurant's dumpling a "historical relic." He had licensed the restaurant's name in Tokyo and Macau and wanted to do the same in Hong Kong.

"What about America?" I asked.

"I'm not sure Americans would like *xiao long bao*," he said.

Frustrated that I had suffered through yet another foul dumpling, I hightailed it back to the city in a taxi.

"What is that?" I asked the driver. A smell like burnt hair filled the car.

"This is the most polluted part of Shanghai," he said. "The property values are lower here than anywhere else in the city." Paint and chemical factories had illegally moved into the district. Many of the residents were recent arrivals from central Shanghai, the driver said. They had been evicted from their old homes, which were being bulldozed for expensive new developments.

Bicycles carrying impossibly large loads plodded down the streets. The apartments were made of cement blocks. People walked on the shoulder of the road because there weren't any sidewalks. We were a world away from the Bund and the French Concession. Now I knew why downtown Shanghainese rarely ventured beyond their cozy bubble of a city.

On a day when neither Little Han nor I was working at Whampoa, we went in search of a restaurant that could make a decent *xiao long bao*. We planned to visit four shops that claimed to serve the best Shanghai soup dumplings. The excursion took us to all corners of the city. We started at Wang Jia Sha, my old neighborhood eatery where my mother had fallen in love with *xiao long bao*. It had undergone a sleek makeover, and now, with glass tabletops and windows that stretched from floor to ceiling, it was unrecognizable from its earlier days as a crowded, dirty canteen. We ended at a dingy shack called Jia Jia, next to the lackluster container ports of the Huangpu River. In between, we stopped into a trendy Taiwanese-owned dumpling eatery called Din Tai Fung, in a neighborhood of *shikumen* row houses that had been converted into a restaurant and bar district. We also crowded into Nanxiang, a sister restaurant to the one that had fed the Cambodian prince. I was skeptical of Nanxiang, but many Shanghainese gave it rave reviews.

We ordered our *xiao long bao* with shredded hairy crab, a seasonal addition that gave the dumplings a buttery flavor and richer texture. It was fall, the time of year when Shanghainese went crazy for the small crabs; Little Han said she was eating three a week. On every street corner, vendors hawked the crawlers from plastic tubs. Fancy restaurants, including the Whampoa Club, devoted tasting menus to the crustaceans. Jereme made drunken hairy crab, the raw meat soaked in sweetened vinegar and Shaoxing wine and swathed in gooey crab roe, which rendered the dish as tender as sashimi yet intensely rich, like caviar. More casual eateries served "crab powder"—a bad translation of "shredded crab." At airports, signs depicted a hairy crab with a diagonal line through it, an ideogram indicating that the item was not allowed in carry-on bags.

At each of our stops, Little Han became spookily quiet when the dumplings arrived, the way my mother or Teacher Jiang would get every so often, thoroughly absorbed by whatever dish had been put in front of them. The young chef picked up the dumplings with her chopsticks, turned them over, and examined them. The *xiao long bao* varied from the size of a ping-pong ball to the circumference of her palm. Skins ranged from spongy to almost see-through. Little Han taught me the finer points of evaluating them. The skin was supposed to be thin but strong; the meat should have a consistency so fine and light that if you shook a raw spoonful of it into a glass of water, it would float like whipped cream. Little Han pinched out the meat with expert precision to examine it.

Up to that day, Little Han had been a Nanxiang devotee. But that changed when the dumpling's starchy skin fell apart as I picked one up, the soup gushing from its side. At around three in the afternoon, we made it to our last eatery, Jia Jia, which we agreed was our favorite. The tender skins practically melted into the soup that it held, a soup so addictively smooth and buttery that it seemed almost criminal to come in such tiny portions. The pork filling crammed inside contained generous bits of crabmeat and roe. We scarfed down all sixteen in the bamboo basket, on top of the dumplings we'd had at our previous stops. We fell into a comatose lull and stared out at a grungy section of the riverfront lined with abandoned factory buildings.

As we sipped tea, Little Han said that she planned to visit Tianjin soon. She had a friend there, a woman who had urged her to move to the northern coastal city and promised to help her get settled. Little Han was getting weary of her job, and she figured that in Tianjin, away from Shanghai's cutthroat competition, she could open a *xiao long bao* stand of her own. She had $20,000 in the bank—a sizable fortune in China— and she planned to use it to get started. There was one catch: her parents wouldn't let her leave home or use her savings to start such a risky undertaking.

I had to agree with her parents. It didn't sound like a good idea. But I sympathized with Little Han. I knew what it was like to be a headstrong, self-assured young woman who wanted the freedom to make her own mistakes.

"As long as I'm living in the same place as my parents, they'll always consider me a child," she wailed.

I could hear the impatience in her voice. She already had much more freedom than the generations that had come before her. Unlike Chairman Wang or Chef Dan, she hadn't been randomly assigned to a job she didn't like. It wasn't by economic necessity that she ended up cooking. She had wished to become a chef because it fulfilled some youthful notion of happiness. But even though her choices had broadened, life could be suffocating. Little Han still wanted more.

As the afternoon sunlight warmed our table, she continued to think aloud about her plans to visit her friend. She'd bring the woman a bouquet of lavender. But with all the restrictions at airports these days, she was afraid she wouldn't be able to fly with the flowers. So, she said, she would take the train, a twelve-hour ride. As the train chugged north, she could hold the fragrant blue gift in her lap the whole way up.

The next time I saw Jereme, after that night in the kitchen, he was in his office, dressed in a button-down oxford shirt and dark slacks. Though he had to walk through the kitchen to get to his office, I rarely saw him pass by; it was as if he magically appeared at his desk like a genie.

His office was about the same size as the dim sum room, and like that alcove, it abutted the kitchen. But instead of making *xiao long bao,* Jereme made grand plans for his restaurant empire. He had tacked on the wall high-definition close-ups of his prized creations that he had taken himself. Celebrity chefs' cookbooks lined one of his shelves. And on his desk sat numerous files that organized a spiraling number of projects.

That morning, he was studying a piece of paper with blue orthogonal lines on it. He had entered a Japanese cooking competition that challenged him to create a dish that resembled the design. He showed me the blueprints for his kitchen in Beijing, the construction of which was already running over budget. Then he turned his attention to another project—the Mandarin Oriental wanted him to compose a new menu for one of their Chinese restaurants, in Bangkok.

When Jereme sat down to design a new menu, he needed two weeks alone, without speaking to anyone. "I keep a notepad by my bed to write down ideas," he said. But now that he was juggling so many projects, it was difficult to find time for solitude. He also insisted on keeping up with the day-to-day affairs of his staff. One morning, as he clicked through the files on his laptop, he showed me an Excel spreadsheet that he was particularly proud of. It showed the breakdown, by weight, of the average amount of crabmeat, roe, claw, and shell his staff picked apart in a day. Sometimes, when he was feeling curious, he went into the kitchen and weighed the day's output of the expensive, coveted ingredients.

"If I check and it doesn't match my records, I know something is wrong," he said.

"But doesn't it vary, depending on the size of the crabs?" I asked.

He shrugged. "Plus or minus ten percent."

It was just a small example of Jereme's suspicious modus operandi. Spending time at Whampoa had shattered my romantic vision of fine dining. Three years before, I'd imagined that Jereme was an artist who would be happy to spend hours philosophizing about eating and his creations. But I discovered he spent more time on his crabmeat spreadsheets than he did thinking about the meaning of his food. I had been naïve: cooking was business, not art.

On my last night at Whampoa, there weren't any sentimental goodbyes. A global consulting company had rented out the entire restaurant, and by night's end, everyone was nearly dropping with exhaustion. To prepare a six-course menu for 109 guests, the chefs had spent the evening bent over a series of trays lined up along the center aisle of stainless steel counters. I remembered something Jereme had said to me—that he wanted to standardize everything at Whampoa so it was "just like McDonald's."

I had been surprised by Jereme's remark. His food seemed to be the antithesis of what came out of a fast-food franchise. But over time I came to realize that, however glamorous the Whampoa Club, however fancy the food looked when it arrived in the dining room, however delicious it tasted, the activity back in the kitchen wasn't so different from what I had seen at the dumpling factory: cooking was repetitive, unglamorous work.

Jereme was a good businessman. He had been brought on board as an employee, but his ambition had gotten him a partnership. He wouldn't disclose his salary or the Whampoa Club's yearly profits. But in the weeks I spent in the restaurant, the dining room filled up nearly every evening, and the lunch rush was often just as hectic. On the last shift of my internship, the chefs buzzed about how the event would bring in $30,000 in several hours.

Recently, Jereme had discovered flaws in the fine dining

business model. For every upscale restaurant, he said, "you need a Jereme Leung." And much as he would have liked it, Jereme Leung couldn't be a ubiquitous presence at a chain of upscale restaurants. Fine dining was exhausting; ingredient costs were high. To expand, he needed something more "scalable." So, he told me, he planned to open a franchise of cafés in China modeled on Starbucks.

But despite focusing on the business, Jereme had to keep up the appearance that he labored in the kitchen. To sell his product, Jereme needed a stage. He needed a dining room and media exposure. He needed his customers to see him as an artist, not a businessman. That's why he had worn his uniform when I had seen him in the kitchen a few nights before. I had heard about a rare performance Jereme had given at the wok. The *jiangbao jiding,* a wok-fried diced chicken, turned out less than stellar. "The sauce was burnt, and the chicken was a black mess," said the person who told me the story. The waitress in charge of delivering the dish wanted to send it back before it went out to the dining room, but changed her mind when she found out that Jereme had made it.

I asked Jereme if focusing on the business had affected his cooking. He blinked. "I've never really thought about it. Alain Ducasse doesn't cook. Jean-Georges cooks very little."

Cooking was, moreover, a ruthless business. Once Jereme had said of Shanghai, "I have never worked in a place with a more complicated human relations environment. It's tougher than Hong Kong, tougher than Singapore, tougher than Malaysia, man." He switched into Mandarin: "The Shanghainese really know how to play the game." Working in Africa might have been worse, but he didn't know. He had never worked there.

Toward the end of my internship, I learned what he'd been alluding to. Just before I started at Whampoa, six chefs had stolen copies of Jereme's signature dishes, quit their jobs, and joined a restaurant across the river that was attempting to mimic Jereme's. They were given a pay raise of $60 each per month—not an insignificant amount, since the chefs made only a few hundred dollars a month at Whampoa. The leader of the mutiny had been one of Jereme's most highly regarded employees.

The restaurant across the river wasn't the only one that had copied his menu. Aside from his often-imitated Shanghai smoked tea eggs, many of his other items had been duplicated around the city. Recently Jereme had decided to take an out-of-town friend to a restaurant he had heard was doing interesting cuisine. When they sat down to eat, he was presented with a copy of his dates stuffed with glutinous rice and pan-seared foie gras. "I have to say, it was not bad," Jereme admitted.

Then there was the credit card scam that some of the waiters were involved in. Jereme disclosed the details reluctantly, long after my internship was over. When a customer paid a bill in cash, the waiters ran their own cards through the system and pocketed the money, thus receiving extra reward points offered by the credit card company. Given how expensive a meal at the Whampoa Club was, the waiters quickly accrued points that could be converted into microwaves and other appliances. Though it seemed victimless to me, Jereme pointed out that he was charged a percentage for the needless credit card transactions. "They're taking advantage of me," he complained. Though he had told the staff that he wouldn't tolerate the scheme, he acknowledged that it could still be continuing without his knowledge.

As a young chef, Jereme had risen in a system where he was expected to play mahjong with his masters and treat them to fancy meals in order to win their favor. In the luxury hotel industry, however, he had been indoctrinated in a regimented, professional environment. When it came to opening a restaurant of his own, he installed all the first-world hardware, but he was still in a developing-world environment. So, naturally, his kitchen had its contradictions: Though Jereme insisted he didn't play favorites, he also pointed out the chefs whom he considered his "disciples." Though he gave his employees two days off a week, the pay was just as lousy as it was in other Chinese restaurants. Though he encouraged his chefs to come up with new dishes, he demanded that they do everything his way.

It shouldn't have been surprising, then, that these contradictions led to disruptions. And, to a large degree, he was helpless to do anything about them. Given China's flimsy laws, it would be difficult, if not impossible, to sue pirating restaurants and former employees. So he channeled his energy into largely useless efforts, like imprinting "Copyright of Jereme Leung" over the mouthwatering photos displayed on his website.

One thing he did have control over, as he had once noted, was the hiring and firing of his staff. When the restaurant first opened, 90 percent of the kitchen had been local, but over the years, he had axed many of his Shanghainese staff, and now the kitchen was only about half local. He had decided to hire fewer chefs with five-star-hotel experience. They had an attitude that was a bit "too European"—too used to an easy life. "You give them a little, and they ask for a lot," he said. He was determined not to make the same mistakes with his upcoming Beijing venture.

On one of my last mornings at Whampoa, I sat with Jereme

in his office as he narrowed his eyes at his staff through the window. "I know how it's going to go in Beijing. If there are any customer complaints, heads will roll. I will have to kill the chicken to warn the monkey," he said, using a Chinese idiom that meant he'd have to punish one person as a warning to the rest. Energized by the idea, Jereme let out a gleeful giggle.

On my last night in Shanghai, Teacher Jiang took me out for authentic local cooking. He had told me that unless I visited Ah Shan's restaurant, I could never say I'd had real Shanghainese food.

Ah Shan's was unlike any Shanghainese eatery I had ever visited. The drafty, warehouse-like space held rickety wood tables, large and round, that looked as if they hadn't been replaced since the restaurant opened in 1983. A series of removable placards hanging from nails listed the day's dishes. Most cost between $1.50 and $3.50.

Ah Shan, the restaurant's owner and chef, was pan-frying a fish head when I poked my nose in the kitchen. He was a rosy-cheeked chef in his sixties with white whiskers covering his upper lip. He went to his rice cooker and proudly removed a mound of eight-treasure rice—sticky grains steamed with eight ingredients, including red bean paste and dates. I sampled the sweet dish, which had the texture of thick pudding, as comforting as Christmas dessert.

"I wake up at two or three in the morning to start cooking the rice," he said. When they closed each night, he and his wife rolled out a bed in the dining room. The restaurant was both their livelihood and their home. "Real Shanghai people eat here. People go away from the city for years and this is the first place they want to come when they are back," he said.

The food was stripped of any Cantonese, northern, or Sichuanese influences. It was plain, pure Shanghai food. Cucumbers were stir-fried with shrimp. Ah Shan red-braised slivers of eel until they were "QQ"—Chinese slang that meant "tender yet chewy." He stir-fried black carp livers with scallions in so much oil they looked as if they were floating in a lake. That was the Shanghai style of eating: greasy food helped *xia fan,* "make the rice go down."

Jiang used the word *tu* to describe the food—it was unsophisticated, uncouth, and earthy. The dishes were served on chipped plates. Rather than having a food caller, Ah Shan had his brother, who was assistant chef, waiter, and cashier rolled into one. Tacked on the wall were dozens of newspaper reviews of the restaurant that sounded curiously genuine. Writers praised its humbleness and honesty. Patrons grew nostalgic for the time before economic reforms, when the Shanghainese had lived more simply, and beef carpaccio and fusion foie gras creations were unknown. Perhaps this was what the fuss over "authentic" food was about—a return to something elemental and unadulterated.

Teacher Jiang had brought another journalist along, and it was fun to see him doing to this journalist what he often did to me. Jiang pushed the plate of black carp livers toward the man, who looked to be my age and was diligently taking notes.

"Try these," Jiang said. The man obliged, picking up his chopsticks and bowing his head toward the livers.

"Aren't they wonderful?" the jolly food critic said. "They're as soft as tofu but they still taste like liver!"

◎ Side Dish 3: Banquet Toasts

After my Whampoa internship, I went to Yangzhou, a pleasant town 150 miles northwest of Shanghai that was known as the birthplace of Huaiyang, one of the "four big" cuisines of China. Though I had eaten Huaiyang cuisine a number of times, I could define it only in sketchy terms: it wasn't spicy or oily, and it didn't use many exotic ingredients. It was known for its intricate knife work—Huaiyang chefs were famous for being able to carve animal and goddess sculptures out of carrots and other vegetables and shred a block of tofu into angel-hair strands.

I also knew that Huaiyang cuisine was fading fast into irrelevancy. In a recent survey conducted in Shanghai, the cooking style had ranked twenty-third in popularity. Many restaurants that still served the food were dilapidated state-owned enterprises. But Shanghai chefs had a reverence for Huaiyang as the basis of modern Shanghainese cooking, and Chef Dan Qin and others suggested I travel to Yangzhou if I wanted to understand what I had been eating in Shanghai all this time.

Like the cuisine it was famous for, Yangzhou recalled earlier times. One of the few remaining cities in China not yet overrun by drab skyscrapers, it was full of low-rise buildings decorated with balconies and stone gates. Canals looped around and through the town, overhung with weeping willows, and lush gardens built around jagged rocks and goldfish ponds dotted the landscape. The sky was often blue rather than hazy gray, as in most Chinese cities.

Though it wasn't particularly important these days, Yangzhou had been one of the biggest, richest cities in the world during the Tang Dynasty (A.D. 618–907). It lay near the southern end of the Grand Canal, which ran from Hangzhou to Beijing. The town had been a hub of the salt and grain trade before the railroad replaced the canal as the main artery.

Master Chen, my guide through Yangzhou's culinary scene, was a senior chef at one of the city's top hotels. He was sweet and unassuming, with a cherubic face, though I discovered he had plenty to brag about. When we had met in 2005, on my first trip to Yangzhou, Master Chen showed me an article that had been published in the *Washington Post*. I had never seen a copy of the *Post* in China, much less one dated 1982. In the article I read that he was one of the first chefs to travel to the United States after the two countries normalized relations, participating in a culinary exhibition in Washington, D.C.

Now that I had returned to Yangzhou with a cooking diploma, Master Chen invited me to a dinner with about a dozen of his friends, seasoned restaurateurs and chefs nearing senior citizenship. He also insisted on treating me to breakfast and lunch the next day. He introduced me as "the intermediate-level, nationally certified chef" and promptly burst into laughter. I was seated between Master Chen and Old Ji, an influential restaurant owner with a face as wrinkled as a prune. "He was once a representative to the People's Congress," Master Chen announced proudly. Old Ji beamed at me and mumbled something in his barely under-standable Yangzhou dialect. When a giant stewed Yangtze River catfish was placed in the center of the table on a lazy Susan, he boldly picked up my chopsticks and went straight for the fish's mouth.

"The maw is the best part," he said, offering me the treasure.

"Oh, no," I protested. "I can't possibly take that with so many other people here."

The lips slipped off the fish's face and onto my plate. I thanked him profusely. After I finished the lips, he picked up my chopsticks again.

"The eyes are the next-best part," he said.

Several dishes teased out the flavor of a central ingredient with delicate seasonings. Tiny raw shrimp were marinated in high-grade rice wine. Chopped celery was blanched, then sprinkled with rice vinegar and oil. Crabs were steamed and served with vinegar.

Master Chen taught me how to eat the crab. First he pulled off the claws and legs and slurped out the meat. Next he turned the crab over and lifted its apron, cracking the shell open like a suitcase. He showed me how to pull off the lungs, which resembled little brown flaps, and find the brown sac of the stomach. "Don't eat the stomach—that's very dirty," he said, discarding it. The banquet room echoed with the sounds of chefs, restaurateurs, and officials from the Huaiyang Cuisine Association sucking and chewing on crabs.

Several of the partially eaten dishes were returned to the kitchen to be folded into new dishes for the table. The half-eaten catfish, stewed in a thick soy sauce gravy, reappeared stir-fried with tofu. The mandarin fish and chicken became a buttery soup in which noodles were boiled. I worried briefly about the hygiene of recycling the ingredients before I let myself be distracted by the rich, wonderful flavors.

I had brought a bottle of French wine as a gift for Master Chen, and he looked touched as he pulled it out of a fancy case lined with silky yellow lace and proudly showed it to everyone at the table. Given the number of cheap knockoffs in the Chinese marketplace, I had my doubts that the wine was really from France, but it was the only imported wine I could find in the city. Master Chen poured the *yangjiu*— "foreigner's wine"—into everyone's glass as they finished their *baijiu*, although more than a few guests tried to refuse. He pointed the bottle in my direction. "Thank you," he mouthed. "Very good."

In the rice terraces I had visited, the locals had simply clinked their glasses. The Huaiyang chefs and restaurateurs toasted with more ceremony. One lifted his tiny glass with his right hand and supported the bottom with the index and middle fingers of his left, as if it held a gallon of liquid. One stood, clinked, smiled, made a toast, chugged the contents, then inverted his cup to show it was empty. The brave chased their *baijiu* with a shot of *yangjiu*.

My toast was lame. "Thank you very, very much for hosting me," I said, in a refrain I would repeat earnestly over the course of the night. The wine tasted like leather. I caught a glimpse of one guest pouring his wine into his neighbor's glass. He covered his empty cup with his palms and pretended to drink the nonexistent contents.

. . .

The next day, a man named Xia Yongguo, with a red base-ball cap that read TORK LIFT 28TH ANNIVERSARY, escorted me around the Huaiyang Cuisine Museum. Though he had a formal title—deputy secretary of the Huaiyang Cuisine Association—his dress was casual, a red T-shirt and red shorts to match his cap, despite the chilly late-autumn weather. He said he was born in the "thirty-sixth year of the Republican era"—in 1947. When I commented that this seemed a strange way of phrasing it, referring to the government in power before the Communists, he quickly added that he was born right before "China's Liberation."

The museum was the first in China dedicated to food, Secretary Xia said as we strolled through the exhibits. They occupied two rooms in a vast Chinese mansion that had once belonged to a salt merchant.

One display case held molded-plastic models of Yangzhou dishes, like the sushi and ramen replicas found in front of Japanese restaurants around the world. Under the glass sat a plate of Yangzhou fried rice, dotted with yellow and green specks reminiscent of scrambled eggs and scallions. On overseas menus, the dish is sometimes called "Yangchow fried rice" or "Yeung chow fried rice." Nearly every version of the dish I had tried contained entirely different ingredients. In a far-fetched attempt to standardize the dish and collect royalties from overseas restaurants, the Huaiyang Cuisine Association had produced a pamphlet that supposedly revealed the official recipe. A copy was exhibited next to the plastic fried rice. The recipe called for rice, eggs, scallops, sea cucumber, scallions, bamboo shoots, ham, and chicken stock. Normally, Secretary Xia said, anyone who wanted the exact proportions of the various ingredients would have to pay for the pamphlet. But since the effort to collect royalties had failed, the association might manage to give me one for free. "We tried to get it trademarked a few years ago, but it was too late," he said with regret.

Imposing charts and diagrams bristling with Chinese characters lined the museum's walls. Secretary Xia and I stopped at a display of pictures of marine life. "Seafood is a central part of Huaiyang cuisine," he narrated. "Our varieties of seafood are very plenti—" He stopped to inspect the pictures more carefully.

"Actually, the shad are virtually extinct now. And we don't see that many anchovies anymore. And there are very few swordfish these days."

He quickly moved on to the next display.

I had first learned the characters for "shad" in my cooking class a year before, when I was still struggling to read Mandarin. The teacher hadn't been as frank as Secretary Xia, mentioning only that shad was one of China's most famous fish, prized for its rich meat, and that it was found in the Yangtze River. The students didn't learn how rare it was.

By the time I arrived for breakfast at Fuchun, the city's well-known steamed-bun restaurant, Master Chen and his friends were already seated. "Shall we start?" Master Chen asked. I picked up my chopsticks and was about to snatch a dumpling when I realized that everyone had frozen with their *baijiu* glasses in midair, waiting for me to raise mine. I put down my chopsticks and hoisted my glass as high as I could.

We ate the Yangzhou version of *xiao long bao*, which were twice as large and soupy as the Shanghai ones. As we sucked melted pork-skin jelly through a straw, Secretary Xia talked about the glories of the region's cuisine.

Yangzhou was still at the forefront of Chinese cuisine, Xia insisted, especially in its emphasis on improving the quality of ingredients. There was a movement toward producing free-range chickens and other environmentally friendly foods, and some of the region's ducks were now raised organically. People were more concerned about pollution and food safety.

"The earth is getting warmer. I bet the evening moon looks rounder in America than it does in China. China is too polluted!" Xia paused. "In the 1950s, if we said something like that, we would have been persecuted. But nowadays no one can deny it."

Everyone at the table began chiming in with opinions on which cities had the worst pollution.

"Tokyo is not as good as Seoul."

"Yangzhou is better than Shanghai."

"Shanghai is better than Beijing."

"Xi'an and Lanzhou are worse than Beijing."

The Huaiyang chefs seemed more aware of environmental issues than chefs I had met elsewhere. Perhaps it had something to do with the cuisine, with its emphasis on simple, fresh ingredients. At lunch, we were served tiny, shelled river shrimp stir-fried with wolfberries. A goose was marinated in rice wine. Cucumbers—the best I had ever tasted— were blanched and soaked in wine, sugar, and soy sauce and cut into thin, crispy slivers. Carp fillets battered with egg whites billowed like clouds. I had struggled to find good *xiao long bao* in Shanghai, but they were delicious everywhere I tried them in Yangzhou.

And so I discovered Huaiyang cuisine. It was the Chinese equivalent of California cuisine à la Alice Waters, the founder of Berkeley's famed Chez Panisse, who had been an evangelist for high-quality, seasonal products. It was nothing like the thick, red, oily, soy-sauced dishes I had eaten with Teacher Jiang in Shanghai. No wonder many Chinese considered modern Shanghainese cuisine bastardized cooking: merging the two cuisines created a fusion-like blend of contradictions.

I also wondered about the future of Huaiyang cuisine. Urban Chinese seemed drawn to restaurants with more kick or novelty—the fiery spices of Sichuanese, the exoticism of minority cuisines from southwest China, the expensive ingredients of Cantonese. With fewer delicacies, lighter sauces, and a focus on simple ingredients, Huaiyang cooking didn't offer much of a wow factor. Perhaps there would be a revival in the future, just as returning to local ingredients had become a hot trend in America. But it didn't look as if it was going to happen anytime soon.

Nevertheless, the Huaiyang Cuisine Association remained hopeful. Secretary Xia was certain I would spread the word about Huaiyang to the rest of the world. "We use simple seasonings and sauces like soy, chicken stock, salt, scallions, and ginger to create a hundred flavors," he said. "Cantonese cuisine has dozens of sauces, but it doesn't have as many flavors."

Secretary Xia seemed to have a bone to pick with the Cantonese. As another round of *baijiu* was poured into everyone's glasses, his face began to glow as brightly as his shirt. He leaned toward me and slurred into my ear: "The Cantonese are very savvy. They know how to market their food. We are more conservative. We spend our time worrying about the simple pleasures of life. But now we know how to spread propaganda about our food." He didn't have to say more. I knew what he meant. All it took was feeding a glutton like me.

He launched into an improvised jingle:

> *Thank you for eating Huaiyang cuisine!*
> *Propagandize Yangzhou! Propagandize China!*
> *When you come to Yangzhou in November*
> *And you see a man in shorts,*
> *Don't think he's got a mental problem,*
> *It's just me!*
> *It's just me!*

YANGZHOU FRIED RICE
(YANGZHOU CHAO FAN)

I never got the Huaiyang Cuisine Association's official recipe for fried rice, but my version is a tasty approximation.

    3   tablespoons vegetable oil
    2   large eggs
    ¼   cup finely chopped scallion
    1   tablespoon minced garlic
    1   medium yellow onion, finely diced
    1   medium carrot, finely diced
    2   tablespoons chicken stock
    2   teaspoons Shaoxing rice wine
    2   teaspoons sesame oil
    1   tablespoon soy sauce
    4   pieces medium dried scallops, soaked in water
        for several hours and finely diced
    1   finely diced fresh sea cucumber (or 3 large fresh
        shiitake mushrooms, finely diced)
    1   cup finely diced winter bamboo shoot
        (preferably fresh, but canned may be used)

4   cups cold cooked rice
1   cup finely diced cooked ham
½   teaspoon salt
½   teaspoon ground white pepper

Add 1 tablespoon of the oil to a wok and place over high heat, swirling the oil to coat the wok thoroughly. When the oil is hot, beat the eggs, add them to the wok, and swirl them around so they cover the bottom and sides in a thin sheet. After 1 minute, flip the egg sheet and cook until firm. Remove it from the heat and cut into small pieces.

Wipe the wok with a towel and add 2 tablespoons of oil. When the wok is hot, add the scallion, garlic, and onions and stir for 1 minute. Add the carrot, chicken stock, rice wine, sesame oil, and soy sauce and cook for 1 to 2 minutes. Add the scallops, sea cucumber, and bamboo shoot and stir for 2 to 3 minutes. Add the rice, ham, salt, and pepper, stirring for 2 minutes. Add the egg, stir for another minute, then remove from the heat and serve.

part four

HUTONG COOKING

15

AFTER BEING AWAY FOR most of the fall, I was glad to return to Beijing. I had never really thought of the capital as home, but pangs of homesickness had plagued me while on the road, and gradually I realized that it wasn't America I was missing, but Beijing. My longing wasn't just for the place, but also the people: Teacher Wang and Chef Zhang, of course, and to my surprise, Craig too, a good friend and fellow journalist whom I had started dating not long before my travels. Not least, I missed the food. I was happy to be back in the land of noodles, dumplings, chilies, and copious amounts of garlic. I was nostalgic for the sight of old Beijingers stockpiling cabbages in their courtyards once the temperature started to drop, and sweet potato vendors reappearing along the roadsides with their coal-filled metal drums, the roasting tubers releasing a sugary scent into the autumn air.

Soon after I returned, one cool evening in late November, I went to visit Chef Zhang. He had opened a new restaurant with a nephew just a few miles from his former noodle stall. I arrived at the little shack, on a dusty road near a set of railroad tracks, at around nine in the evening, just as the pair were cleaning up and getting ready to close.

"You've gotten skinny!" Zhang said as I stepped through the sliding door.

It wasn't a compliment—skinny was bad, it meant I looked unhealthy—but I knew Zhang was merely showing his concern.

"Really?" I said. "I think I gained weight."

"Well, you shouldn't get any skinnier," he said. "You look tired."

"I wish I were skinnier," I said.

"Why? I want to be fat like him," Zhang said, pointing to his nephew, who was sweeping chopsticks, napkins, and bits of food from the cement floor. His slumped posture emphasized his roly-poly figure, which was complimented by a round, chubby face. Zhang was just as bony as the last time I had seen him, and his face was more gaunt, making his large, round eyes stand out. "No matter how much I eat, I can't get fat," Zhang said, shaking his head.

He brought out some food to plump me up. The restaurant had a more extensive menu than the noodle shop, with a roster of Shanxi specialties I had never heard of. He handed me a bamboo basket full of steamed noodles called *kao laolao*. I dipped the wide, penne-shaped oat pasta in a spicy vinegar and coriander sauce and felt the coarse, chewy noodles go down with a kick. A friend of Zhang's brought the vinegar directly from Shanxi, since Zhang didn't trust the supply in Beijing. "It could be fake," he said.

Chilly air seeped through the thin walls of the restaurant, and I was shivering despite my down coat. I did my best to hide how cold I was, not wanting Zhang to lose face. But I was curious why he had chosen this location, after a month of bicycling around the city looking at prospective spots.

"To be truthful, this is not my ideal situation," Zhang said.

"We chose this place because the rent was cheap. It's the best we could do without borrowing money or having to go into business with anyone." Old Wang, his former coworker, still wanted to open a restaurant with him, but Zhang decided that, as a migrant, he'd be at a disadvantage if he partnered with a Beijinger, who would always have the law on his side.

A single poster with Chinese characters decorated the place, which the landlord had put up before Zhang arrived. Zhang explained that it was a passage from Buddhist scripture. The landlord had run a vegetarian restaurant, but the cuisine wasn't very popular in the area, and the place had gone out of business.

"The landlord is a strange kind of vegetarian," Zhang said. "He doesn't even eat eggs. A bald guy, but he's definitely no monk. He is sly, very slick. Before I took over the place, he counted all the plastic cups and plants he had left and recorded it in his notebook. He wanted me to keep the phone in here so he could overcharge me for the phone bill. He warned me not to scuff up the floor." The floor was made of cement.

Zhang asked about Craig. I told him that though I had been away for most of the fall, we had managed to see each other once in a while, and we were spending more time together now that I was back in Beijing.

"I have a question for you," he said. "Is he part Chinese?"

Zhang had met Craig, who had never been mistaken for Chinese, with his blue eyes and wavy light brown hair. "No," I said, laughing. "Why do you ask?"

"I don't know," Zhang said. "There is something about him that makes him seem Chinese. He doesn't carry himself like a foreigner."

That much was true — Craig had lived in China for years and had none of the brash arrogance of many expatriates. He spoke Mandarin well. He was unassuming, a quality that Chinese prized.

I asked Zhang how his family was doing, especially his son.

"He's good. You two get along quite well. I don't have a close relationship with him. He thinks of me as a stranger since I've been away for so long. The only extended stay I've had in Shanxi since I left was during SARS." The restaurant had laid him off during the epidemic, so he'd spent three months at home. But without any income, the period had been more stressful than relaxing.

I asked if he'd been afraid of contracting SARS.

"I wasn't afraid then. I'm a traditional Chinese person. I believe in fate. There are certain things you can't control. I wasn't lying in bed waiting for it to strike me. Some things are destined. Like getting to know you. It was fate that I started that first noodle shop—if I hadn't started that shop, the girls wouldn't have brought you over and I wouldn't have gotten to know you. Some things you can anticipate and control. Other things you can't."

After chatting for an hour or so, I stood up to go. I knew that Zhang didn't mind sticking around in the restaurant rather than falling asleep in an uncomfortable temporary bed —he had moved again, after his family returned to Shanxi— but my body was beginning to feel numb from the cold. So I thanked him for the noodles, waved goodbye, and flagged a taxi back to my apartment, where I knew the heat would be cranked up high.

A few months later, Zhang and his nephew closed the eatery. He had decided to leave Beijing. A friend of his had gotten him a job managing a canteen in a small town eight hours away by train. The canteen fed managers who were overseeing the construction of one of China's massive highways. Zhang had decided he could no longer afford to feed other people all day and still worry about how he would feed himself. He needed the security of a rice bowl. But he vowed that eventually he'd come back to Beijing.

I spent the winter getting to know Craig better and discovered that if there was one thing that disqualified him from being Chinese—even more than his looks—it was that he was not interested in food. Not long into our courtship, he once admitted, "I wish a pill were invented that could take the place of eating." He thought of food as mere fuel, and ate quickly, preferring to be doing anything rather than eating, which he thought of as a waste of time. He had a terrible morning Nescafé habit; he considered it too much trouble to grind coffee and put it through a French press or a drip. He judged a Chinese restaurant on how well it made kung pao chicken, the most basic of dishes. While I enjoyed tuna, oysters, and beef raw or rare, he insisted on eating them well done and chided me for taking chances with my health. (I supposed he had a point—we were in China, after all. Even so, I defiantly continued my risky behavior.)

Somehow we each managed to overlook what we regarded as the other's character flaw. We spent a lot of time in restaurants anyway, as most courting couples do, the settings of our dates covering an interesting cross-section of Beijing's dining scene. Our first official date was in a lofty, stylish Sichuan eatery in the art district, filled with long-haired painters and musicians. Another evening we joined friends at a grungy neighborhood restaurant that served the lamb- and potato-heavy cuisine of the Uighurs. We ventured to the outskirts of the city to eat in a converted peasant's house set in an apple orchard. I could not recall what I had eaten at any of these restaurants, because for the first time in a long while, it didn't matter. I was too busy falling in love.

In the spring, Chairman Wang and her husband moved from their drab apartment block to a single-level home down a *hutong*. The two-room residence occupied one corner of a large property built around a courtyard in central Beijing. Traditional courtyard dwellings consisted of four long rooms, constructed of brick, wooden beams, and gray roof tiles. The rooms formed a rectangular fortress around a garden, and a gate provided the only access to the outside world. Though originally intended for wealthy, single families, many courtyard homes were now broken up into smaller units. The Wangs shared their compound with nine other families. As was often the case, the courtyard had been swallowed up by a cluster of shanties. From the gate, a narrow, dusty path wound around the courtyard, leading to the doors of the various residences.

The Wangs' place had an interesting history. Mr. Wang's working-class parents had managed to scrape together enough money to buy the rooms before the Communists came to power, and Mr. Wang had grown up there. During the Cultural Revolution, one of the rooms was taken away from the family as part of a campaign to redistribute wealth. Mr. Wang and his mother continued to occupy the one remaining room after his father died. But after Mr. Wang married, there wasn't enough space for all three of them, so the young couple spent their first year of marriage in Mr. Wang's office.

When Chairman Wang gave birth to her son, she decided to write a letter to the vice mayor of Beijing. She told the vice mayor that her husband was a teacher. They had an infant. An office wasn't a proper environment to raise a child. Could the government kindly see about returning the second room?

The next week, Chairman Wang received a letter saying that the government was looking into the matter. A month later, a functionary visited her in her husband's office to ask a few questions. A few more months passed, and then the government notified the Wangs that they would get the room back. ("Something like that would never happen these days!" Chairman Wang said as she recounted the story.) The new family moved into the reclaimed room, where they stayed until they moved into their cement block apartment in the early 1980s. The building was a major step up: they went from two rooms and a public toilet to three spacious rooms, a separate kitchen, and their own private flush toilet.

After Mr. Wang's mother passed away, several of his relatives had occupied the *hutong* dwelling. Chairman Wang and her husband had thought about moving back to the *hutong* in recent years; they wanted to give their modern apartment to their son, who was going to marry his girlfriend soon, but they were too polite to ask their relatives squatting in the *hutong* to leave. I couldn't imagine that happening in America: letting relatives occupy my home indefinitely, the rent coming in the form of an occasional case of apples. Finally, though, the relatives found a new place, and they turned over the home to Chairman Wang and her husband. Now that the Wangs had moved back, they were converting a small attached patio into a third room with a bathroom and a spacious kitchen. (*Hutong* residents who were lucky enough to be able to afford their own toilets had no qualms about putting them next to their kitchens, given the limited plumbing in the neighborhoods.)

The renovation was all-consuming. The trunk of an old date tree that grew outside the Wangs' home was pushing up against the wall. To cut it down, they needed permission from the appropriate government bureau, which required bribing officials with expensive cartons of cigarettes. They hired a team of migrant workers to cut down the tree and lay the brick for several new walls. The Wangs had initially planned to add a second story, but when the construction team began building it, a neighbor threatened to complain to the local authorities, second floors in the area being technically illegal. To smooth over relations with the neighbor, the Wangs offered *baijiu*. Chairman Wang had chosen the bathroom and kitchen fixtures herself, at one of the interior-decorating superstores in Beijing. (Subcontractors couldn't be trusted, she said.) The Wangs stayed glued to the site throughout the renovation, standing over the workers to make sure they didn't steal any bricks or paint. Three months, five thousand dollars, ten cartons of cigarettes, and a dozen bottles of *baijiu* later, the Wangs' bathroom and kitchen sparkled with new appliances. A shiny silver refrigerator and washing machine sat next to a two-burner stove. The numerous cupboards would hold all the snacks and staples that Chairman Wang desired. A round dining table fit comfortably in the room. The bathroom had bright, white-tiled walls and white porcelain fixtures.

It was nothing like her previous kitchen—the shabby, bare-bones room where Chairman Wang had squatted at her cutting board, working by the light of that single bulb hanging from a nail in the wall. I congratulated her on the work.

"It's *hai xing*," she said. Not bad. Her wide cheeks pushed up against her big glasses as she smiled.

The Dragon Boat Festival had come around again, so one

morning I joined Chairman Wang to make *zongzi,* the rice dumplings wrapped in reed leaves. We got straight to work. But rather than use her new countertops, she set us up next to the door, on little stools at a midget sized table like the one in her previous kitchen. I didn't ask why. I was too busy wrestling with the dumplings. I could collect the grains of sticky rice in a reed-leaf envelope, and I managed to fold the leaves over so that they completely covered the rice. But I was as clumsy as ever at bundling the package, rice grains leaking from its sides as the string unraveled.

"You're hopeless!" Chairman Wang yelled with glee.

To cook for someone, one on one, is to let that person into your life in an intimate way. I was hesitant to cook for Craig, even as winter gave way to spring and we got more comfortable with each other. A previous relationship had dissolved soon after an evening when I had cooked, and although I knew the meal had been just fine, I sensed that the guy had taken my cooking, with its domestic connotations, as a sign that we were getting too serious. Not wanting to put those expectations on my relationship with Craig, I had reserved my cooking for Teacher Wang, old friends, and the occasional out-of-town guest.

Craig, on the other hand, didn't have the same kitchen intimacy issues and approached the task of cooking fearlessly in my presence, chopping, seasoning, and tossing with reckless abandon. One evening, he invited me over for a rendition of his "one-pot cooking." I wasn't quite sure how he did it, since I was too horrified to watch, but he managed to make pasta with vegetables and shredded processed cheese without dirtying more than a single pot. Another evening, he proposed making chicken and mushrooms. It would be simple, he said. He would buy a can of Campbell's cream of mushroom soup from the foreign supermarket, mix it with a few chicken breasts, and bake the concoction in my toaster oven (regular-size ovens being a rarity in China). He defended the recipe as an American classic. I pointed out that we weren't in America, and laid down the law: we were never going to have such an abomination in my home.

I found myself getting nervous every time Craig offered to cook. Whenever he did, I changed the subject and, after a few minutes, casually brought up the exciting new Sichuanese, Shanghainese, or even Russian restaurant around the corner. But I sensed that Craig was getting suspicious. Why didn't I cook? Had I really been in cooking school and interning in restaurants all this time? In any case, I couldn't resist cooking for my handsome new boyfriend, who had an athletic six-foot build, cornflower-blue eyes, and a disarming smile. So I ventured into the kitchen. I began with a Western standard, spaghetti carbonara. A few weeks later, I tentatively offered a sample of the leftover *mapo* tofu I knew I had nailed the evening before. "This is delicious!" he said as he demolished the dish within seconds. "The best *mapo* tofu I've ever had."

Suddenly bursting with confidence, I agreed to cook an elaborate Chinese New Year's dinner at his place for six of our friends. They were supposed to show up at eight o'clock. Around three in the afternoon, as I looked at all the groceries we had amassed in the kitchen, I panicked. "Why don't we just go out for dinner?" I suggested. Surely, somewhere in town was a restaurant that wasn't completely booked.

Craig reassured me that all would be well as he joined me in the kitchen. He began washing and chopping as if he were the Tasmanian Devil, and we managed to pull it off. The eight-course meal began with a collection of dishes I had learned over the previous year and a half. I made the sweet-and-sour pork and dry-roasted Sichuan string beans I'd learned in cooking school, the red-braised eggplant from Shanghai, and minced pork in lettuce wraps, a dish I improvised. We ended with dumplings, as important to Chinese New Year as turkey is on Thanksgiving. The only disaster was the smoked duck, a recipe of Chef Dan's, which I overcooked. It came out of the toaster oven looking like a piece of wood—well done, just the way Craig liked it. That was the best part of dating a nonfoodie: he enjoyed whatever I cooked. At midnight, we climbed up to Craig's roof and watched an amazing 360-degree fireworks display. Shimmers of red, yellow, and blue streaked the evening sky, accompanied by blasts rivaling the sounds of Baghdad: the work of the collective citizenry of Beijing, who cherish their right to buy and launch as many fireworks as they want.

"THE BEST" *MAPO* TOFU

- 2 tablespoons vegetable oil
- ¼ pound ground beef
- 2 tablespoons minced leek or scallion
- 1 teaspoon minced ginger
- ¼ cup broadbean paste
- 2 tablespoons soy sauce
- ¼ teaspoon salt
- ½ teaspoon sugar
- ½ cup water
- 1 package firm tofu, cut into ¼-inch cubes
- ½ teaspoon ground Sichuan peppercorns
- 1 tablespoon Shaoxing rice wine

Add the oil to a wok and place over high heat. When the oil is hot, add the beef, breaking it into small pieces and stirring until it begins to brown. Add the following ingredients, stirring for a minute between each addition: leek and ginger, broadbean paste, soy sauce, rice wine, salt, and sugar. Add the water, reduce the heat to medium, and simmer for 4 to 5 minutes. Add the tofu, raise the heat to high, and stir for another 2 to 3 minutes. Sprinkle in the ground Sichuan peppercorns and remove from the heat. Serve immediately.

In the summer, after months of delay, Jereme Leung's Whampoa Club Beijing opened. The restaurant was located in a tiny green oasis of a park amid the new glass-and-steel structures of Beijing's characterless financial district. A decade earlier, it had been one of the old courtyard neighborhoods, but when the area was zoned to become the capital's financial hub, everything was bulldozed. Someone high up in the government must have had second thoughts about the loss of so much old architecture, however, because three models of traditional courtyard houses were rebuilt, using some of the original tiles and bricks. Jereme's restaurant occupied one of these courtyard buildings.

Craig and I went for a meal one balmy evening. I had decided not to take him to the Whampoa Club when he'd visited me in Shanghai because he disliked fine dining; it was a waste of both time and money. "Why does it take two hours to eat?" he once asked at a fancy restaurant, as bored and antsy as a ten-year-old. On another night out on the town, his eyeballs bulged from their sockets as he perused the menu: "Seven-fifty for a beer? Are you kidding me?" The very fact that I was able to drag him to the Whampoa Club Beijing was a sign of progress.

The courtyard contained private dining rooms and a bar with black lacquered walls, dark leather seats, and teal-colored silk throw pillows. A dimly lit hallway led to an underground foyer adorned with black birdcages of various sizes and expanded into a large dining room with a stunning feature: a glass-bottom goldfish pond sat atop the room, allowing sunlight in to illuminate the diners below, the rays glinting off metal baubles that dangled from the ceiling. It was an ingenious use of a basement space, but as the sun set, the baubles resembled icicles and the place grew dark and cavernous. I also noticed that certain details, like the blue horsehair-upholstered walls in the bathroom and seats covered with strips of shiny leather, mimicked the design of Jean-Georges's restaurant in Shanghai. I doubted that the touches were accidental.

We were seated in a booth that was separated from the main dining room by blue velvet curtains and furnished with a baroque black chandelier and Bose speakers. As we opened the menus, I silently noted another improvement in Craig: he didn't so much as blink at the prices. A few seconds later, however, he guffawed.

"Did you see this?" he asked. Pointing to the front page of the menu, he read: "Masterminded by gifted and enterprising founder-chef Jereme Leung, Whampoa Club Shanghai is acclaimed and acknowledged as one of the best fine dining restaurants in the world. Beijing proudly joins the ranks of Whampoa Club, the epitome of unsurpassed Chinese cuisine."

"Yeah, he is pretty arrogant," I said.

"Arrogant?" Craig said. "Or insecure?"

A waiter dressed in a black robe and a yellow belt came by. I ordered a $75 tasting menu, and Craig ordered à la carte: wok-fried chicken and tea-smoked pressed bean curd. Despite the gorgeous setting—or perhaps because of it—I was wary about the food. Just weeks earlier, Jereme had confessed that he had not been able to devote enough time to the recipes. I had recently heard about the incident of the lackluster wok-fried chicken. And it was clear that Jereme had put a lot of money and effort into the interior design of the place. I wondered if it was going to be another example of a flashy Beijing restaurant that placed style over substance.

A long white plate arrived at the table bearing a colorful collection of appetizers arranged in neat, bite-sized bundles. I was surprised at how well the flavors worked together. Velvety foie gras terrine—the dish that Jereme once despised—was embedded in a delicate bean curd roll. A tower of pickled vegetables—the very ingredient that Jereme wouldn't touch before living in mainland China—had a terrific, crunchy mouthfeel. A roll of *youmai,* a leafy Chinese vegetable, drizzled with a spicy sesame dressing felt smooth yet fiery on my tongue. Shreds of potato were finely julienned, showing off a chef's deft knife work. A small mound of cabbage had a wasabi kick—a kick that was part of Jereme's trademark style.

Jereme had the chefs send out a tasting menu for Craig as well, and the chicken and tofu he ordered, though quite tasty, were left to cool as we enjoyed the procession of dishes that followed. It was the first time I had seen Craig at peace in a fine dining restaurant; he even seemed to be enjoying the meal.

The appetizers alone erased my doubts about Jereme's abilities. The courses that followed were a little uneven. A hot-and-sour soup with lobster squandered a perfectly good piece of seafood. A roasted lamb tasted heavy and out of place. But the presentation of each dish was so spectacular that it almost made up for the unbalanced flavors. The final dishes redeemed the meal. A delicious cod, baked in a sheath of spring onions, was tender and flavorful. Jereme's interpretation of traditional Beijing *zhajiang* noodles came with a colorful display of julienned pink radish, carrot, cucumber, and bits of corn with a sauce spiked with Sichuan peppercorns. The pan-fried northern-style dumpling, folded with such care that it bore the intricate pattern of a leaf, was so juicy it sprayed an arc of broth across the table when Craig bit into it. Dessert was a triumph—a flaky almond tart served with a traditional northern almond tea in a delicate clay teapot. All in all, it was the most refined northern cuisine I had ever sampled—never mind that some people would argue that northern food wasn't meant to be refined.

At the end of the meal, Jereme stopped by to say hello. He was wearing a white shirt and jeans and looked more relaxed than I had seen him in a long time. I complimented him on the food.

"It's the Year of the Pig," he said as he sat down in our booth. "I was born in the Year of the Pig. The Chinese believe that your year is either a very bad or a very good year."

(His comment was prescient. In the first half year of business, the restaurant faced another mutiny of chefs, and his investors canned his café concept before his luck turned the following year, the Year of the Rat, as more guests filled his dining room and the restaurant began to win awards.)

Jereme added that he liked Beijing. The northern chefs were more laid-back and less greedy than the Shanghainese. "Like tonight," he said. "I'm going to take them out for lamb skewers, and they're all happy. If it were in Shanghai, they would ask, 'Who's going to pay the taxi fare?'"

We chatted for an hour or so. He was getting ready to go on a beach holiday in Southeast Asia. He missed the slower pace of the tropics. He mentioned that he didn't think he was going to stay in mainland China forever—eventually, someday, after he built his restaurant empire, he would retire somewhere nice. By the time Craig and I got up to leave, it was past eleven o'clock. As Jereme escorted us out, he asked one of the hostesses where the chefs were. He hadn't eaten all day and was ready to take them out.

"They've all gone home," she said.

Craig and I grew more serious, and at the beginning of the summer, I moved in with him. He lived in a two-story apartment within a three-floor compound, a rarity in a *hutong* neighborhood. An architecture professor at Tsinghua University had designed the compound as a project in sustainable living in the late 1980s, with the idea of replicating it throughout the neighborhood, but the plan had dissolved once the first few compounds had been built. Though the apartment had modern amenities like a shower, toilets (two), and a kitchen, it was rustic enough to remind us of the surrounding area. The power went out when too many people ran their air conditioners in the summer, and we would have to bundle up in layers come winter, with the wheezing radiators barely putting out heat.

But I loved the location. Nestled in the alleys, the apartment was shielded from the sound of honking cars. Craig and I took sunset strolls to Houhai Lake. The Hualian Cooking School, where I had started all my adventures, was within walking distance. The biggest perk of all: Chairman Wang was just a five-minute bicycle ride away.

"*Gongxi, gongxi!* Congratulations!" she cried when I told her that Craig and I were engaged. Her gray perm had grown out, and she had cut her thick straight strands into a practical crew cut for the summer. She was wearing the same flowery blue *qipao* that she had worn to my dumpling party almost a year ago.

Most other Chinese had been unimpressed when I told them my news. They seemed to expect that from the moment you started dating someone, you were going to get married. In a society where many women referred to their boyfriends as their *laogong*, their husbands, the idea of an official engagement aroused suspicion: what had Craig and I been up to before, when we weren't engaged?

But that afternoon was not meant for celebration. We had serious business. Chairman Wang and I set out to scour the neighborhood for a space. We planned to open a cooking school. Since I had begun cooking, I had gotten dozens of requests from friends and acquaintances—where could they learn to cook? We had held a few trial classes in friends' kitchens, and now we decided it was time to look for a proper place to hold classes.

Just down the alley from my home, a landlord showed us a *hutong* apartment that seemed perfect: two large rooms that looked out onto a garden, with a small alcove that could, he said, become a bathroom.

The next day I called a contractor, who came to give me an estimate for the renovations.

"You can't put a bathroom there," he said.

"Why not?" I asked.

"A toilet has to be within a few meters of a septic tank," he said. He led me around the perimeter of the building to show me the nearest septic tank, which was marked by a manhole with the character *wu,* or dirt, on it. The manhole was nowhere near where we wanted to install the toilet. We restarted our search for a cooking school space.

After that incident, as I walked through the alleys each day, I noticed for the first time all the manhole covers hinting at the invisible network that lay beneath the surface: *xinxi*, for telecommunications; *biao*, for meters; *zha*, for switches. They were a good reminder of all the subtle details of China that I had yet to learn.

One morning as Chairman Wang and I cooked in her kitchen, I complained about the lack of meat, vegetables, and fruit for sale in our neighborhood. All the mom-and-pop convenience stores sold were instant noodles, soft drinks, and processed cookies.

"Have you been to the market near the Drum Tower?" Chairman Wang asked. I hadn't. She needed to remedy that right away. As soon as we finished steaming a batch of corncakes, we got on our bicycles and pedaled to the market. She had an elegant way of riding, her feet pointed slightly outward, her posture straight. She breezed down a *hutong* and turned onto the main street, gracefully weaving through traffic as I trailed behind.

We arrived at a set of warehouses that were hidden behind a row of restaurants. The airy space, as large as a hangar, was filled with vendors and shoppers bargaining over meat, fruit, vegetables, and grains. Heaps of rice and nuts took up one room, and shoppers nearby inspected peaches and *hami* melon, a fruit the size and texture of a watermelon but with the flavor and look of a cantaloupe. Butchers sharpened their knives as they took orders for pork belly and pig's feet. The market was twice as large as the one Chairman Wang had showed me near her previous apartment. I couldn't believe that I had lived in the neighborhood for a couple of months and hadn't come across the market until then.

But that was what living in a *hutong* was like: discoveries, like the manhole covers, came gradually, giving me a fuller picture of Beijing's traditional life if only I was willing to be patient. One morning, I strolled out of our apartment and stumbled upon a woman on a bicycle pulling a blanket-covered wagon behind her. "*Mai-cai-yah!*" she shouted with the pipes of a Mongolian throat singer. She lifted the blanket to reveal a load of cauliflower, tomatoes, cucumbers, and onions. So that was why the nearby stores lacked fresh produce. I began to hear her reliable call every morning and evening. "*Da mi! Da mi! Bai mian fen!*" another vendor shouted, hauling a load of rice and white flour on his wagon. The knife sharpener rode by reliably every other day around four in the afternoon: "*Mo dao! Mo dao!*" he shouted, rattling his knives against a metal board and pausing long enough in our compound for me to bring him my cleaver.

I was glad to discover the vendors on bicycles before it was too late. So much of Beijing had changed in the race to become a modern, world-class capital that the city was losing its character and its food traditions. Common street snacks like *jianbing*, a salty crepe brushed with chili sauce and sprinkled with coriander, were disappearing. Old noodle shops were being replaced by cafés that served bland spaghetti bolognese. Food markets like the ones that Chairman Wang had showed me were becoming rarities as Wal-Marts and Carrefours popped up around the city.

And all the while, as Beijing moved forward, I was regressing. Coming to China in the first place was, in my parents' view, a counterintuitive thing to do. My mother's family had fled from China—they had worked so hard to get out—why did I want to go back? They didn't understand why I hadn't wanted a stable job in the United States, and why I had taken my chances working in a Communist country. And now, after becoming a cook, I had taken another step back, into one of the decrepit old neighborhoods that were being bulldozed by the block.

Through all the regressing, I had found equilibrium. Finally, after living in China for seven years, I was content. I had been blessed with good food, friends, love, and a small understanding of history that made me appreciate what it all meant. My parents had forgiven me for moving to China and approved of my passion for cooking after I visited them in California and made them a seven-course meal that filled their house with the scent of oil, chilies, and peppercorns. In Beijing, I had found a home with Craig, and Chairman Wang was just around the corner. As for the numerous Chinese I had come across on my cooking journey who were still struggling to find their way— Chef Zhang, Little Han, the waitress Qin, the dumpling workers, among others—I wondered how things would eventually turn out for them.

Then I remembered something that Chairman Wang had told me once: "*Suan, tian, ku, la.* Sour, sweet, bitter, spicy. In my life, I've tasted them all." Her life had been more bitter than sweet, more sour than spicy, but things were *hai xing*, not bad, and that gave me hope for the rest.

A Note on Recipes

Unless otherwise noted, the only special equipment needed to prepare the recipes in this book are a cleaver and a wok—a fourteen-inch cast-iron model is ideal.

Chinese dishes are typically served family style. When cooking dinner for four, prepare four to six dishes, and accompany them with a batch of dumplings or noodles (see recipes) or steamed rice (two cups of raw rice steamed in a rice cooker should be enough).

Throughout the recipes, scallions (primarily used in southern China) and leeks (used in northern China) may be used interchangeably. Most Chinese use soybean oil for cooking; feel free to substitute canola or another mild-flavored vegetable oil. Olive oil may also be used, except in those recipes that call for deep-frying.

Sources

To supplement Chairman Wang's memories of her childhood and the Cultural Revolution, I used information from Jasper Becker's *Hungry Ghosts* (New York: Henry Holt, 1966), Roderick MacFarquhar and Michael Schoenhals's *Mao's Last Revolution* (Cambridge: Belknap Press of Harvard University Press, 2006), and Jonathan Spence's *The Search for Modern China* (New York: W. W. Norton, 1990).

For information on Chinese festivals used in Part One and Part Three, I relied on S. C. Moey's *Chinese Feasts and Festivals: A Cookbook* (Singapore: Periplus Editions, 2006).

To gain a better understanding of monosodium glutamate, I read "Could MSG Make a Comeback?" by Sara Dickerman (*Slate*, May 3, 2006); "It's All a Matter of Taste" by Fuchsia Dunlop (*Financial Times*, August 6, 2005); Karl Gerth's *China Made: Consumer Culture and the Creation of the Nation* (Cambridge: Harvard University Asia Center, 2004); "MSG: A Common Flavor Enhancer" by Michelle Meadows (*FDA Consumer Magazine*, January–February 2003); "In China, MSG Is No Headache, It's a New Treat" by Elisabeth Rosenthal (*New York Times*, December 14, 2000); "A Short History of MSG" by Jordan Sand (*Gastronomica*, Fall 2005, 38–49); "The Safety Evaluation of Monosodium Glutamate" by Ronald Walker and John R. Lupien (*Journal of Nutrition*, volume 130, 2000); and

"Rivers Run Black, and Chinese Die of Cancer" by Jim Yardley (*New York Times*, September 12, 2004).

I relied on the following articles for background and history of Shanghai: "Shanghai's Culinary Great Leap Forward" by Geoff Dyer (*Financial Times*, August 19–20, 2006); Irene Corbally Kuhn, "Shanghai: The Vintage Years," in *Endless Feasts: Sixty Years of Writing from "Gourmet,"* edited and with an introduction by Ruth Reichl (New York: Modern Library, 2002); Jereme Leung, *New Shanghai Cuisine* (Singapore: Marshall Cavendish, 2005); "Shanghai Cuisine's History" (*Shanghai Food and Culture Newsletter*, issue 4, August 28, 2006, in Mandarin); and "From Slop to Sophistication, Shanghai Undergoes a Culinary Transformation" by Stan Sesser (*Wall Street Journal*, October 18, 2002).

Acknowledgments

I couldn't have written this book without having met such wonderful people through my cooking endeavors, the most important being Wang Guizhen. She is the most patient teacher I've ever had, and I value her candor and wisdom on just about everything.

I was enormously lucky to have been befriended by Zhang Aifeng, who graciously let me into his kitchen and taught me much about migrant lives. Qin and the Cai triplets welcomed me into their dormitory and showed me how waitresses lived. I am grateful to Jereme Leung for the experience of interning in his Shanghai operation and for introducing me to his gifted Whampoa chefs who taught me how to finesse my cooking. Thank you also to Jiang Liyang in Shanghai for sharing his food knowledge with me over the years. Dan Qin and Takashi Miyanaka were kind to allow me into Yin. I would also like to thank the Liao family in Guangxi and Chen Chunsong in Yangzhou for their warm hospitality. Thank you also to Gregory Veeck and Karl Gerth for sharing their research and helping me form ideas that made their way into the text. Wang Xin, Jonathan Ansfield, and Jimmy Hu accompanied me on some of the more adventurous meals that appeared in this book.

Leslie T. Chang was tremendously helpful in giving me advice on my book proposal and connecting me with her agent, now our agent. Leslie, you are the best friend a writer could hope for. It was a true bonding experience to go through the book process together and to be able to share the pains and joys of writing (and everything else) along the way.

I appreciate the help of Jenny Chio, who painstakingly read my first draft and gave me constructive comments that improved the book. Peter Hessler has encouraged me from my early days of writing in China and has been a wonderful mentor. I'd like to thank Dan Brody, Mark Leong, and Chandler Jurinka for their early support as well. (Chandler, the pep talk you gave me at Bodhi somehow stuck with me through the whole project.) I appreciate the friendship of Tara Duffy, Maureen Fan, and Sherisse Pham, who shared their sympathetic ears, almond croissants, and chocolate chip cookies. Jin Ling thoroughly fact-checked the manuscript.

William Jefferson Foster introduced me to the waitresses. Thank you to Odilon Couzin and Yuen Chan for arranging the dumpling internship. Handel Lee and Michelle Wan helped me set up the internship in Shanghai. Tom Pattinson at *Time Out Beijing* allowed me to write brutally honest restaurant reviews and gave me space for a monthly column on eating. E. C. Liu and P. T. Black hosted me in their respective homes in Shanghai.

I am also grateful to Kirsten Wegner, Charlotte Sector, and Mary Helen Kelt for their friendship overseas and for giving me encouragement and advice along the way. Christopher McCormick, Shai Oster, and Robin Moroney read the endless drafts of my book proposal. Jeremy Brown read later drafts of the book and offered his historical expertise. I am indebted to Yun-Yi Goh, Rachel Ruderman, and Judy Hu for their longtime friendship and encouragement. Victor Shih was my Peet's Coffee mule. Sig and Mary Gissler in New York have also been very supportive throughout.

Brook Larmer, Evan Osnos, Oliver August, Peter Goodman,

Deanna Fei, and Elise Goodman offered sound advice and helped me choose the perfect agent and publisher.

Chris Calhoun believed in me from the very beginning, and I am indebted to him for his enthusiasm for the project. Becky Saletan was endlessly dedicated and meticulous in her editing. On top of her usual duties as publisher, she worked her magic on the manuscript, striking the right balance by giving the book a polish without changing the content and meaning of anything I wrote. She was the right editor for this book, and I'm grateful to Chris for leading me to her. I also appreciate the help of Laurence Cooper, Sarah Melnyk, and Thomas Bouman.

I am also grateful for the friendship and support of my future parents-in-law, Caroline and Dave Simons, with whom Craig and I have shared many adventures in Beijing. My grandparents in Taipei shared stories about food that inspired sections of this book, and they've fed me countless meals over the years.

Thank you to my family: Mom, Dad, and James. I'm grateful for the support—financial, emotional, and otherwise—they've given me since I was born. My parents' relationship with food, from their love of oysters to their obsession with fish head (and James's affection for vinegar and ketchup), helped me develop my own adventurous taste buds. I appreciate James's enthusiasm for my writing, even when it came to tough subjects like eating dog.

Finally, I'd like to thank my soon-to-be husband, Craig Simons, for hauling the manuscript everywhere we went, from Wudaohe to Whole Foods in New York. Craig, it was invaluable to have you with me on this journey: to listen, advise, edit, read, cook, eat, drink, and celebrate. It takes an extraordinary man to have been as understanding, wise, and compassionate as you, and I love you very much.